REMOTE SENSING APPLICATIONS IN CROP MANAGEMENT:

MONITORING, DISEASE DETECTION, AND PRODUCTIVITY OPTIMIZATION

REMOTE SENSING APPLICATIONS IN CROP MANAGEMENT:
MONITORING, DISEASE DETECTION, AND PRODUCTIVITY OPTIMIZATION

Edited by

Dr. Mukesh Laichattiwar

Delve PUBLISHING

www.delvepublishing.com

Remote Sensing Applications in Crop Management: Monitoring, Disease Detection, and Productivity Optimization

Dr. Mukesh Laichattiwar

Delve Publishing
4164 Lakeshore Road
Burlington, ON L7L 1A4
Canada
www.delvepublishing.com

Email: orders@arclereducation.com

ISBN: 978-1-77956-399-6 (Hardcover)

ABOUT THE AUTHOR

Dr. Mukesh Laichattiwar, Ph.D., currently serves as an Assistant Professor at the College of Agriculture, Parul University, Vadodara, Gujarat. He earned his Ph.D. and M.Sc. (Agri.) in Agricultural Entomology from the Institute of Agricultural Sciences, Banaras Hindu University (BHU), Varanasi, Uttar Pradesh, and his B.Sc. (Agri.) from the College of Agriculture, Ambajogai, affiliated with Vasantrao Naik Marathwada Krishi Vidyapeeth, Parbhani, Maharashtra.

Before joining Parul University, Dr. Laichattiwar served as an Assistant Professor at the College of Agriculture, Naigaon (Bz.), Nanded, Maharashtra, from 2018 to 2019. Dr. Laichattiwar has made significant contributions to the field of agricultural entomology, evidenced by his extensive publication record, which includes 20 research articles, 5 edited books, 10 book chapters/proceedings, 5 popular articles, and 15 abstracts.

A recognized contributor to the academic community, Dr. Laichattiwar has been honoured with several awards, including a UGC Fellowship for his Ph.D. research and a Best Poster Presentation award in 2014. Additionally, he qualified the ICAR-NET in 2013. He is also actively engaged in the academic community, having attended 15 national and international seminars. Dr. Laichattiwar is a life member of four professional societies, underscoring his commitment to advancing agricultural sciences.

TABLE OF CONTENTS

LIST OF FIGURES

LIST OF TABLES

PREFACE

Remote sensing in agriculture utilizes advanced technology to observe and measure crops from a distance. This technology includes satellites, drones, and sensors that collect data about fields. By analyzing this data, farmers can monitor crop health, soil conditions, and water usage, as well as detect diseases early. This approach enables precise decision-making to improve crop yield and efficiency, ensuring better management and sustainability of agricultural practices.

The book is divided into eight chapters. Chapter 1, "Remote Sensing as a Management and Monitoring Tool," introduces the fundamental concepts of remote sensing and its application in agricultural management, setting the stage for understanding how technology can enhance traditional farming practices. Chapter 2, "Introduction to Crop Management," provides an overview of crop management principles, emphasizing the importance of integrating modern technology into these practices.

Chapter 3, "Remote Sensing in Crop Disease Detection and Management," explores specific techniques and case studies that demonstrate the effectiveness of remote sensing in identifying and managing crop diseases. Chapter 4, "Remote Sensing for Insect Pest and Disease Management," extends this discussion, illustrating how these technologies can be employed to address a broader range of agricultural challenges.

Chapter 5, "Role of Images in Remote Sensing," examines the significance of imagery in remote sensing, detailing how different types of images and sensors contribute to accurate disease detection. Chapter 6, "Estimation of Crop Water Productivity via Remote Sensing," focuses on how remote sensing can be used to monitor and enhance water-use efficiency in agriculture. Chapter 7, "Remote Sensing of Rice Crop Areas," offers a specialized look at the application of these technologies in rice cultivation, one of the world's most important staple crops. The book concludes with Chapter 8, revisiting the initial concepts with a deeper understanding of remote sensing's role in agriculture.

This book serves as a vital resource for anyone involved in agriculture, from researchers and students to farmers and policymakers. By leveraging the power of remote sensing, we can drive forward a new era of precision agriculture that promises to enhance productivity, sustainability, and resilience in the face of global agricultural challenges.

—Author

Chapter **1**

REMOTE SENSING AS A MANAGEMENT AND MONITORING TOOL

Contents

1.1. INTRODUCTION

Remote sensing is a powerful tool that provides new insights into agriculture, particularly in the area of crop disease detection and management. By utilizing satellites, drones, and other advanced technologies, it is possible to gather detailed images and data about crops from a distance. This information is crucial for the early detection of crop diseases, allowing for timely intervention before the diseases spread and cause significant damage. Early detection enables farmers to take effective measures to manage and control diseases, resulting in healthier crops and improved yields. The ability to monitor large areas efficiently and accurately makes remote sensing an invaluable resource for modern agriculture (Aase & Siddoway, 1981).

Remote sensing technology has revolutionized agriculture management and monitoring by providing valuable information on crop health, soil conditions, weather patterns, and overall land management. The advancements in satellite, aerial, and ground-based remote sensing, have increased the use of reflectance data in agriculture. Conventional agriculture systems often fail to optimize resource utilization and soil productivity. However, taking the aid of remote sensing techniques helps boost the productivity potential of the soil and optimize the inputs (Qian et al., 2019).

In the past century, agriculture has undergone a significant transition from conventional methods to technology-intensive practices. This shift has been driven by the need to meet food needs because although the world's population has doubled and food demand has tripled since 1960, global food demand has been met with only a 30% increase in cultivated land and intensification of the inputs like fertilizers, pesticides which harm the environment. According to projections, the world population is expected to grow to almost 10 billion by 2050, a significant increase from the almost 8 billion recorded in 2019. Additionally, there is a projected 56% increase in global food demand. That is why it is important to ensure that agricultural production is sustainable and does not have negative impacts on the environment. This can be achieved by making efficient use of available resources.

Remote sensing has proven to be incredibly useful in a wide range of fields, including resource mapping, detecting land use patterns, investigating geo-hydrological conditions, and monitoring droughts and floods. Utilizing remote sensing has also benefited agro-meteorology in a variety of ways, such as determining the temperatures of clouds, land, precipitation, radiation, soil moisture, crop yield, and so on. In addition, this technology is also useful for monitoring disease and pest incidence. Remote sensing, when used alongside crop simulation models, is a valuable tool for predicting crop yield (Parry et al., 1995). This chapter highlights the importance of remote sensing in agriculture.

The upcoming section will delve into the theoretical background of applying remote sensing in agriculture.

1.2. THEORETICAL BACKGROUND

The basis of remote sensing applications in agriculture lies in the interaction between electromagnetic radiation and agricultural features such as vegetation, soil, and water. Understanding these interactions is crucial for extracting meaningful information from remotely sensed data and utilizing it for agricultural purposes. The information provided by remote sensing is transmitted through EM radiation, traveling at the speed of light through a vacuum as waves of different wavelengths (Liu et al., 2016). A section of the electromagnetic spectrum that spans the visible, infrared (IR), and microwave bands is used in remote sensing. Across the electromagnetic spectrum, the spectral signatures reflect the distinct reflectance patterns of various agricultural features. These signatures are determined by the absorption and reflection properties of the features. They are used to distinguish vegetation from bare soil, water, and other similar features based on the responses of the targets to these wavelength regions (*see* Figure 1.1). To prevent data loss and enable simple access for researchers globally, the spectral signature graph is kept in a digital database within the spectral signatures library (Singh et al., 2007). Through the analysis of spectral signatures, remote sensing can effectively differentiate various land cover classes and track their transformations over time. By comparing the spectral signatures of vegetation, soil, and other agricultural components with known spectral libraries, RS can identify and discriminate between different land cover classes and monitor their changes over time. Another approach to tapping information from RS is the Vegetation Index (VI). Vegetation Indices (Vis) are quite simple and effective algorithms for quantitative and qualitative evaluations of vegetation cover, visor, and growth dynamics, among other applications (Ban et al., 2008).

1.2.1. Vegetation Indices (Vis)

The biological and physical characteristics of plants can be differentiated spectrally via Vis termed as unitless radiometric quantities. They are computed in terms of differences/ratios of two or multiple bands of SWIR, NIR, and visible wavelengths. The usefulness of Vis can be determined by its strong correlation with the physical characteristics of plants, while being less affected by factors that make interpreting remotely sensed data difficult, such as relief, atmospheric conditions, viewing angles, soil background, non-photosynthesizing plant elements, and lighting conditions (Xu et al., 2008).

Ratio Vegetation Index (RVI) is a type of vegetation index. With its high sensitivity to vegetation and strong relationship with the plant biomass, the RVI (ratio of red to IR wavelength) is widely used for monitoring and estimating green biomass, especially in areas with dense vegetation coverage. The Normalized Difference Vegetation Index or NDVI is commonly used in crop management and remote sensing. It is calculated as the ratio of the difference to the sum of the NIR and Red reflectance. The utilization of NDVI time series data has proven to be highly effective in a wide range of applications. These include detecting global changes, monitoring crop growth, and predicting yields, studying phenology, monitoring drought and desertification, assessing wildfires, and modeling biogeochemical and climatic factors (Haile et al., 2019).

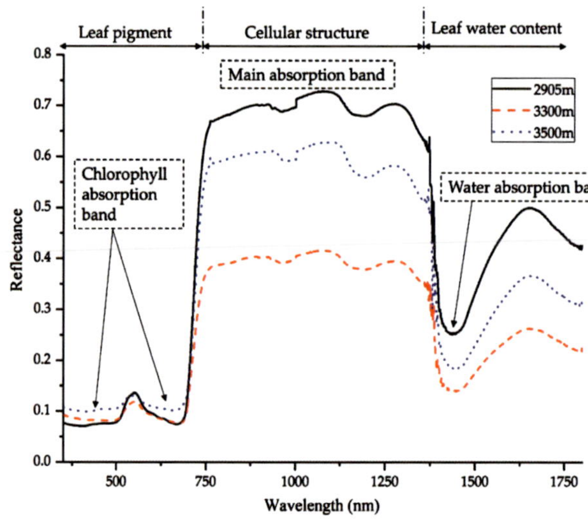

Figure 1.1. *Spectral characteristics of Quercus aquifolioides leaves.*

Source: Zhu, J., He, W., Yao, J., Yu, Q., Xu, C., Huang, H., & Mhae, B., Jandug, C. (2020). Spectral reflectance characteristics and chlorophyll content estimation model of Quercus aquifolioides leaves at different altitudes in Sejila Mountain. Applied Sciences, 10(10), 3636. https://doi.org/10.3390/app10103636 (Creative Commons License).

Several research has been carried out to develop additional indices that account for the background effects of soil and atmospheric conditions on spectral measurements. SAVI, a well-known vegetation index, effectively accounts for the influence of soil background on vegetation cover sensed remotely.

The Visible Atmospheric Resistant Index, or VARI, is an index that is comparable to this (Vrešak et al., 2016). This index significantly reduces the impact of the atmosphere. Several indices have been developed to assess variations in the SWIR and NIR wavelength reflectance, indicating the presence of water deficiency in plants. Examples include the Water Index or WI and Shortwave Infrared Water Stress Index or SIWSI. To consider the connection between thermal properties and water stress of plants, researchers have developed indices like Stress Index Water or SI and Deficit Index or WDI. Over 100 Vis have been reported from multi-spectral imageries (Torres et al., 2019). Several vegetation indices (Vis) utilized for specific agricultural purposes are documented in Table 1.1.

Table 1.1. *Description of Spectral Parameters*

Spectral Parameters	Definition
Red edge position (REP)	The wavelength position corresponds to RES.
Red edge slope (RES)	The largest first-order derivative value in the red edge (680~750 nm).
Blue edge position (BEP)	The wavelength position corresponds to BES.
Blue edge slope (BES)	The largest first-order derivative value in the blue edge (490~530 nm).
Yellow edge position (YEP)	The wavelength position corresponds to YES.
Yellow edge slope (YES)	The largest first-order derivative value in the yellow edge (560~640 nm).
Red valley position (RVP)	The wavelength position corresponds to RRV.
Reflectance of Red Valley (RRV)	Minimum reflectance in the wavelength range of 640~700 nm
Green peak position (GPP)	The wavelength position corresponds to RGP.
Reflectance of green peak (RGP)	Maximum reflectance in the wavelength range of 510~580 nm
Reflectance of water stress band (RWSB)	Maximum reflectivity in the wavelength range of 1550~1750 nm.
Red edge area (REA)	Sum of the first derivative in red edge (680~750 nm).
Yellow edge area (YEA)	The sum of the first derivative in yellow edge (490~530 nm).
Blue edge area (BEA)	Sum of the first derivative in blue edge (560~640 nm).
Leaf chlorophyll index (LCI)	$(R_{850}-R_{710})/(R_{850}-R_{680})$.

Source: Zhu, J., He, W., Yao, J., Yu, Q., Xu, C., Huang, H., & Mhae, B., Jandug, C. (2020). Spectral reflectance characteristics and chlorophyll content estimation model of Quercus aquifolioides leaves at different altitudes in Sejila Mountain. Applied Sciences, 10(10), 3636. https://doi.org/10.3390/app10103636.

1.3. REMOTE SENSING PLATFORMS DEVELOPMENT

There are three distinct platforms for applying remote sensing technology in agriculture:

i. Ground-based;

ii. Airborne; and

iii. Space-borne satellites.

Each platform has its advantages and disadvantages, and the method chosen depends on the purpose and the scale of the survey.

Using the ground-based remote sensing method, field-scale sensors are utilized to monitor abiotic and biotic crop stresses. These sensors can be handheld or mounted on machines. This method offers significant advantages, including improved spectral; spatial; and temporal resolution. Nevertheless, the scale; efficiency; and labor re□uired in this techni□ue pose limitations on its applicability, restricting it to smaller areas. On the other hand, airborne and satellite-based remote sensing methods are more suitable for larger areas (Spanic et al., 2010).

Airborne remote sensing is the next platform, which utilizes manned aircraft alongside UAVs or drones to provide timely pictorial crop inventory. The selection of equipment is determined by the allocation of funds. A typical UAV consists of navigation and communication systems along with multiple sensors onboard. Unmanned aerial vehicles (UAVs) are lightweight, cost-effective, and operate at lower speeds, making them a viable alternative to manned aircraft. There are two different kinds of (UAVs): "fixed wing" and "rotary wing." Each UAV has a different flying time based on the weight of its cargo. The former, with a payload of light HD cameras (300 gms), can fly at high speed for two hours without the need for a launcher or runway, while the latter can hover over an object for a shorter amount of time (15–25 minutes) because of its larger payload weight and higher battery power consumption. Table 1.2 lists different UAVs used for agricultural purposes, specifically for monitoring the health of crops remotely (Nielsen et al., 2014).

The UAV platform is a highly cost-effective and flexible solution, providing a great alternative to traditional airborne and satellite methods. High-resolution imagery is produced by UAVs due to their ability to cover a target quickly and repeatedly at various times and altitudes. As a result, the UAV can detect differences among the plants throughout the field a comprehensive analysis of the applications of UAVs in agriculture has been extensively examined by multiple experts.

The ultimate platform is space-borne remote sensing, which is classified based on the orbit and the time of a satellite (Xue et al., 2019). The data captured by this satellite is suitable for much bigger geographical regions, enabling the monitoring of crops on a global and regional scale. In addition, it can be utilized on a local scale to provide valuable yield forecasts, classification, mapping, and coverage of crops. One drawback of this method is its susceptibility to potential cost inefficiency; and susceptibility to atmospheric noise and meteorological conditions. In addition, the revisit time of satellites, such as Landsat (16 days) and SPOT (26 days), is relatively long, which can restrict their usefulness in agriculture, especially for tasks like irrigation and nutrient scheduling (Femenias et al., 2021).

Table 1.2. *Agricultural UAV Platform Usage*

Application		Sensor		Processing method				
				Photogrammetry	Machine Learning	Vegetation Indices	ML & VIs	Other
Weed mapping	12.5%	RGB	73%	12.50%	62.50%	12.50%	0.00%	12.50%
		Multispectral	27%	0.00%	100.00%	0.00%	0.00%	0.00%
Monitor growth	65.6%	RGB	30%	42.11%	31.58%	10.53%	0.00%	15.79%
		Multispectral	59%	13.89%	25.00%	38.89%	8.33%	13.89%
		Hyperspectral	6%	25.00%	25.00%	25.00%	25.00%	0.00%
		Thermal	5%	0.00%	33.33%	66.67%	0.00%	0.00%
Monitor health	6.3%	RGB	20%	0.00%	100.00%	0.00%	0.00%	0.00%
		Multispectral	80%	0.00%	25.00%	25.00%	50.00%	0.00%
Irrigation management	5.2%	Multispectral	60%	0.00%	0.00%	33.33%	66.67%	0.00%
		Thermal	40%	0.00%	50.00%	0.00%	0.00%	50.00%

Source: Tsouros, D. C., Bibi, S., & Sarigiannidis, P. G. (2019). A review on UAV-based applications for precision agriculture. Information, 10(11), 349. https://doi.org/10.3390/info10110349.

Some satellites offer data for free, while others provide commercial solutions. One such commercial satellite is Pleiades-1, you can access high-resolution images that provide a temporal resolution of one day. Some of the widely used satellites for acquiring hyperspectral imagery include Landsat-8 and Sentinel, which offer free solutions, and QuickBird, which is a commercial option. Despite the higher temporal resolution of Landsat-8 compared to QuickBird, it offers multi-spectral images in 11 different bands (Dong et al., 2020). There are three missions of the widely used satellite called Sentinel, namely Sentinel-1; Sentinel-2; and Sentinel-3.

Because of Sentinel-1's high revisit frequency with a 6-day interval, insensitivity to cloud cover, and rapid reaction of SAR to water content and vegetal structure, it has significant promise for mapping crops. With the utilization of the dual-polarization and multi-temporal features of SAR data, Sentinel-1 is capable of generating highly accurate crop maps, achieving an impressive accuracy rate of 85%. One of the most commonly used products for agriculture in this mission is the IW Level-1 GRDH product. It requires a thorough pre-processing procedure that includes 1). Calibration; 2). Coregistration; 3). Multi-looking; and 4). Geocoding. Sentinel-2 has a lot of promise for monitoring agricultural fields both locally and globally because of its global coverage, high spatial resolution (10–20 m), high revisit frequency of 5 days, and integration with the Landsat series. Sentinel-3's operational feature suggests that its primary design drivers are widely accessible data products with significantly □uicker delivery times (Shah et al., 2019).

1.4. REMOTE SENSING APPLICATIONS IN AGRICULTURE

1.4.1. Evaluation of Biomass

The spectral as well as structural properties of the target can be acquired by RS at various spatial and temporal scales, hence making it the best possible method for huge-area biomass evaluation. Accurate estimation of AGB (above-ground biomass) is crucial for effective crop growth monitoring. Currently, AGB evaluation is done using ground-based and/or satellite remote sensing techniques. Due to its labor-intensive and time-consuming nature, ground-based remote sensing may not be scalable. On the other hand, satellite remote sensing might not provide the necessary level of data resolution to be effectively used in precision agriculture (Ma et al., 2020).

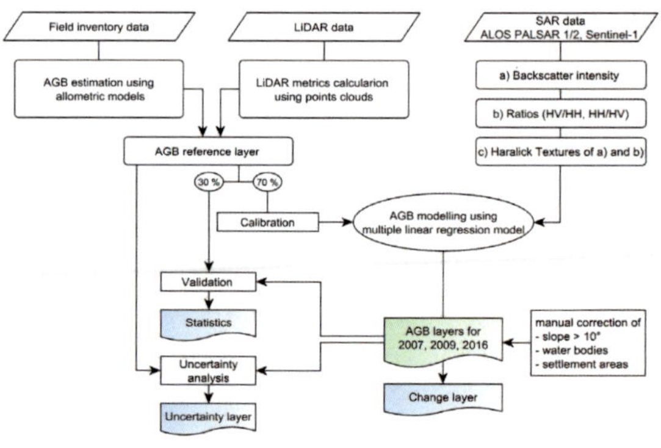

Figure 1.2. *Flow chart of the AGB map processing steps.*

Source: Berninger, A., Lohberger, S., Stängel, M., & Siegert, F. (2018). SAR-based estimation of above-ground biomass and its changes in tropical forests of Kalimantan using L- and C-band. Remote Sensing, 10(6), 831. https://doi.org/10.3390/rs10060831 (Creative Commons License).

LiDAR (light detection and ranging) is another technique commonly used to estimate biomass. LiDAR data has gained significant attention due to its reliability in estimating biomass, as it overcomes the data saturation issue of Landsat. LiDAR offers a highly accurate crop surface model because of its high spatial resolution. This allows for precise estimation of biomass by measuring plant height. However, it is important to note that LiDAR is only suitable for small areas due to the extensive computational resources needed for processing. UAVs and lightweight sensors have bridged the gap between terrestrial and satellite, offering exciting possibilities for precision agriculture. In recent times, there has been a growing interest in using UAV-centered remote sensing methods to monitor crops, with a particular focus on physical plant factors such as plant height. It has been discussed that plant height, which may be acquired via canopy surface models or CSMs (canopy surface models) at various temporal scales, is a reliable metric for estimating biomass. In addition to CSMs, RGB, and multi-and hyper-spectral images collected with UAVs have been used with CSMs to assess biomass (Mahlein et al., 2012) (Figure 1.2).

Three-dimensional (3D) point clouds are created from successively overlapped 2D images that were obtained by a UAV via the Structure from Motion or SfM algorithm. These point clouds present new possibilities for the acquisition of crop surfaces. SfM, a computer-based technology, generates 3D geometry through the use of image matching and repeated bundle adjustment techniques. Derived from CSMs, 3D point clouds contain valuable information about the vertical distribution of crop canopies. This information can be harnessed for various crop monitoring purposes, such as measuring plant height, predicting yield, and estimating biomass (Feng et al., 2022).

A type of digital elevation model (DEM) derived from point clouds is the triangulated irregular network (TIN), which represents the surface using a sequence of continuous, non-overlapping, asymmetrical triangles through the Delaunay triangulation algorithm. TIN can be efficiently utilized for the extraction of crown volume. TIN, as opposed to gridded DEM, may more effectively and precisely display surface structure details without the interruption procedure]. In addition, the data analyzed by TIN provides a higher level of detail and goes beyond just measuring plant height. This indicates that there is a wealth of additional structural information that can be utilized for assessing AGB. Using multispectral data in combination with structural features improves the accuracy of assessing AGB compared to using either of them alone. This improvement is further enhanced by incorporating meteorological features (Alisaac et al., 2018).

1.4.2. Prediction or Estimation of Yield

The yield estimation of the crop is of great significance not only to farmers but also to government bodies and policy framers to boost agricultural productivity and spot the abiotic and biotic threats affecting crop productivity. The site-specific yield estimation of crops helps to discern the spatial variability of crop vigor within the plot and thereby, aids in optimizing management strategies and minimizing crop threats (Liu et al., 2015).

The forecasting of crop yield using RS technology is principally based on the empirical/statistical relationships existing between yield and Vis. Among the various indices that have been developed, NDVI is widely used to assess biomass, developmental phases, crop conditions, and ultimately yield. It is closely linked to vegetation vigor and has been proven to be highly effective. Some studies have found that the green leaf area index or LAI of certain crops exhibits a non-linear asymptotic relationship with NDVI. Various LAI values reflect different intercepted wavelengths that directly impact biomass generation, which in turn provides insights into crop yield. In addition, the NDVI contains a significant amount of information regarding precipitation and can effectively explain the variability in grain production, such as in wheat (Guo et al., 2021).

However, the empirical models that estimate yield tapping RS data are limited to those regions only where they have been calibrated. Yet they are used commonly because they are less data intensive and easy to employ at the regional level. For an accurate estimation of crop yield at the regional level, many researchers have integrated spectral data with crop growth simulation models, but these models are very data-intensive encompassing soil properties, crop management practices, crop parameters, and agro-meteorological data (Chang et al., 2015).

The Ground-based RS apparatuses have been expended successfully for forecasting the yield of many crops-rice wheat. The determination of accurate yield predictions in heterogeneous agricultural landscapes using satellite data is dependent upon the resolution of sensors as well as other exogenous factors. The freely available sensors like the AVHRR (Advanced Very High-Resolution Radiometer) are appropriate for detecting vegetation changes, on account of the high temporal resolution, however, the spatial resolution of these sensors is low of the order of >8 km which reduces its efficiency in estimating the crop yield in heterogeneous land settings. Moreover, most of the farmer's land holdings are smaller than the pixel size of the sensor, hence making AVHRR difficult to use for yield estimation. The significant advances in the field of RS technology gave rise to successor sensors such as MODIS, an improvement over AVHRR, especially in terms of spectral and spatial resolutions. MODIS was explicitly aimed at land-related studies allowing foliage monitoring at the regional level.

The yield estimates using MODIS data are more precise than AVHRR data when applied to comparatively homogeneous fields. The MODIS has an improved spectral resolution (36 spectral bands); however, 250 m spatial resolution is still coarser for heterogeneous settings. On account of these limitations, sensors with improved spectral, temporal, and spatial resolutions like the Landsat series were framed, which offered an apt substitute to MODIS and AVHRR for crop yield prediction. The Landsat 8 OLI having a spatial resolution of 30 m is such an example. However, it suffers from the major drawback of fairly long temporal resolution, having a periodicity of 16 days, confining the number of observations taken in a crop growing season, which is further fueled by cloud cover limiting the successful sensing. Despite this limitation, Landsat series (e.g., TM and ETM+) with revisit frequency akin to Landsat 8 OLI have been reported to provide fairly accurate yield estimation, when applied to homogeneous agricultural systems Landsat 8 OLI sensors have been even found suitable for heterogeneous African cropping systems for capturing phonological stages of maize (Liao et al., 2018).

1.4.3. The Global Food Demand

Globally per capita food demand is strongly linked with per capita Gross Domestic Product (GDP; Figure 1.3). For example, the wealthiest nations (represented by group A, such as the US) consume around 8,000 kcal·day^{-1}, while groups C and D (consisting of Brazilian and Indonesian people, respectively) consume 4,000 kcal·day^{-1}. While considering this and presuming that both GDP as well as global populace will endure proliferating in the future, the earlier trend of intensely escalating food pressure is projected to persist for 3–4 decades envisaging that crop caloric/protein demand per head will double between 2005 and 2050 (Barbedo et al., 2015).

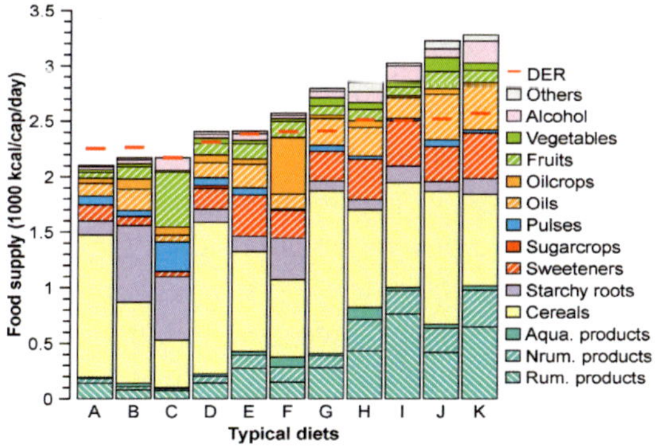

Figure 1.3. *The annual relationship between the per capita GDP of various economic groupings and the crop caloric demand per capita.*

Source: Pradhan, P., & Kropp, J. P. (2020). The interplay between diets, health, and climate change. Sustainability, 12(9), 3878. https://doi.org/10.3390/su12093878 (Creative Commons License).

The data obtained through remote sensing technology has immense potential in monitoring food demand and quality, providing timely, comprehensive, cost-effective, and regular information. From the data obtained from current remote sensing satellites, one can evaluate the crop production and acreage. In addition, it is possible to detect disturbances, conditions of stress, and phenology of crops. With the future bringing unpredictable weather patterns, it is crucial to closely monitor agricultural production for risk assessment. This monitoring needs to be done at temporal and spatial scales and can only be achieved with the help of remote sensing techniques (Zhang et al., 2019).

The two remote sensing methods—optical RS-based surface reflectance and microwave RS-based backscattering—are used for predicting as well as mapping. A nation must have a dependable and timely food forecast system before the crop is harvested, regardless of whether remote sensing or ground-based methods are used. Utilizing remote sensing techniques, experts in crop management and remote sensing can provide valuable food production forecasts. These forecasts can assist governments, decision-makers, and policymakers in developing appropriate strategies. For example, they can help quantify the volume of food that needs to be imported in the event of a deficiency or the volume of food that can be exported in case of a surplus. This information can also help governments procure food from other nations at more affordable rates without revealing any details about potential future food shortages (Whetton et al., 2018).

1.4.4. Detection of Salinity Stress

The salinity of the soil has an adverse effect on the quality of the soil and water, resulting in land degradation that hinders crop growth and ultimately reduces crop yield. RS technology has revolutionized crop monitoring and mapping, allowing us to not only track crop growth but also identify crop stresses caused by saline conditions.

Remote sensing technology can assist in identifying salt traits that are visible on the soil surface, such as white crust, which directly indicates soil salinity. Another method for determining soil salinity is through the use of various indicators, such as the halophytes presence and crops that are tolerant to high salt levels. Several studies have used multispectral and hyperspectral sensors to map and monitor soil salinity areas (Huang et al., 2019). Based on research findings, hyperspectral sensors are more effective than multispectral sensors in detecting salt characteristics on soil surfaces and distinguishing between halophytes and non-halophytes. This is mainly due to the superior spatial resolution of the hyperspectral sensors (Mahlein, 2016).

1.4.5. Drought Assessment and Monitoring

Drought is a multifaceted hydro-meteorological phenomenon that occurs when there is a prolonged decrease in precipitation, leading to a decrease in soil moisture. Drought can take on various forms and is categorized into different types, like hydrological; agricultural; or meteorological drought. One commonly employed method for assessing and tracking drought is through the use of a drought index. A comprehensive inventory of drought indices and indicators, along with their practicality, has been presented by the Global Water Partnership (GWP); and the World Meteorological Organization (WMO). Precipitation plays a crucial role in the characterization and occurrence of drought through the use of drought indices (Whetton et al., 2018). Precipitation data exhibits significant temporal and spatial variability, making it challenging to always accurately interpolate. The effectiveness of drought indices relies heavily on the availability of rain gauging stations in a given area, which poses a challenge for assessing drought in regions with just a handful of rain gauges. The advancements in remote sensing techniques have made it possible to obtain precipitation data with varying temporal and spatial resolutions. These techniques have been widely utilized for drought forecasting on a global scale. When there is a lack of rainfall, known as a meteorological drought, it leads to a shortage of soil moisture, which is called an agricultural drought. Remote sensing is a valuable tool for monitoring agricultural drought. It allows us to assess soil moisture status and evaluate various vegetation indices such as Soil Adjusted Vegetation Index (SAVI); Vegetation Condition Index (VCI); Vegetation Health Index (VHI); Enhanced Vegetation Index (EVI); Evaporative Stress Index (ESI); and NDVI (Zheng et al., 2018).

1.4.6. Soil Moisture

Understanding soil moisture levels is crucial for managing water resources, predicting natural upheavals, and mitigating the impact of natural disasters such as dust storms, erosion, floods, and droughts. However, accurately measuring soil moisture (in situ) can be expensive and time-consuming, as it requires frequent soil sampling to track varying vicissitudes in the moisture levels. The remote sensing technique provides a synoptic view of broad areas coupled with temporal and spatial fluctuations, making it a better alternative to the standard ways of determining soil moisture (Barbedo et al., 2017). The near-gamma radiation, microwave, TIR, IR, and visible; IR; TIR; micro; and near-gamma radiation are used to remotely sense the moisture content of the soil. However, due to the ability to penetrate the ground and the year-round effectiveness of passive as well as microwave radiations, these techniques have proven to be very promising. Cloud cover poses a significant obstacle to the effectiveness of rest techniques (Mahlein et al., 2019). While remote sensing data obtained through microwave radiations shows promise, its interpretation can be challenging due to factors such as soil properties, surface roughness, and vegetal cover (Huang et al., 2020).

1.4.7. Change Detection

Understanding and implementing sustainable land use practices is crucial for ensuring food security. Understanding and monitoring changes in land use and land cover (LULC) is crucial for ensuring food security in a region, as agricultural land is often converted to other uses.

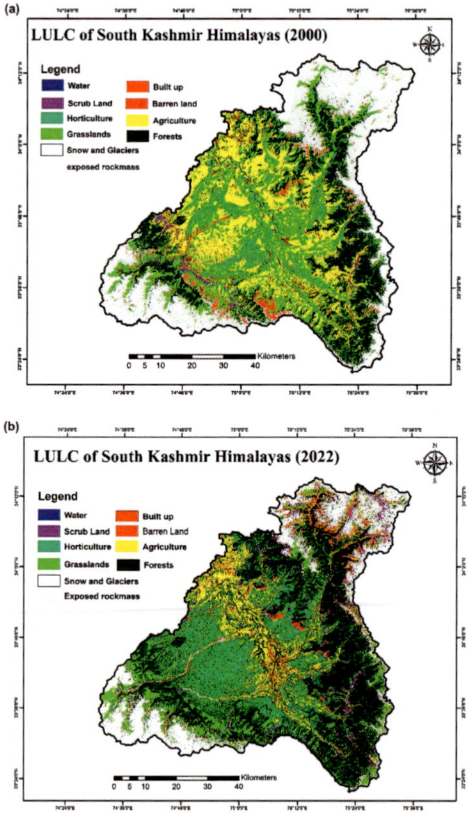

Figure 1.4. *Changes in land use and cover over time in Southern Kashmir Himalaya, J&K, India.*

Source: Mushtaq, S., Saleem, S., Ahmed, R., et al. (2024). Spatio-temporal analysis of land use land cover changes in the South Kashmir region of North–Western Himalayas using Landsat data. Discover Geoscience, 2, 37. https://doi.org/10.1007/s44288-024-00031-3 (Creative Commons License).

Mapping LULC trends and changes has proven to be a beneficial use of satellite-based remote-sensed data. Studying changes in land use and land cover can provide valuable information for policymakers to develop effective land use policies for a specific region. Figure 1.4 showcases a notable case of land use and land cover change study in the Southern Kashmir Himalaya, J&K, India. It highlights the significant transformation of agricultural land into horticultural land over time (Delwiche et al., 2019).

1.5. SUMMARY

The chapter discusses the important role that remote sensing plays in the agricultural industry. Farmers and other agricultural stakeholders can maximize resource usage, increase productivity, and promote sustainability by utilizing remote sensing techniques like satellite images, sensor-based data collection, and aerial photography. Remote sensing data is incredibly efficient in empowering agricultural actors to address global issues like climate change; degradation of land; and loss of biodiversity. This chapter offers a thorough examination of several possible applications of remote sensing in agriculture and emphasizes its importance in enhancing sustainability; resource efficiency; and overall productivity. It has investigated the potential applications of remote sensing technologies in the areas of estimation of biomass; estimation of yield; management of global food demand; detection of salinity stress; monitoring the drought conditions; assessment of soil moisture content; and detection of changes in the agricultural landscapes. Every application has been thoroughly examined, taking into account the advantages and obstacles it presents, emphasizing the importance of remote sensing in agricultural operations. Researchers have successfully utilized RS technology to measure crop parameters and predict crop yield. Remote sensing data can accurately measure vegetation anomalies such as salinity stress and drought conditions, which can have a significant impact on crop yield. This data produced by remote sensing is crucial to the forecasting of global food demand, which aids governments and policymakers in creating plans for regulating agricultural production and obtaining food supplies in the event of a shortage.

The combination of remote sensing and cutting-edge data analysis techni□ues, like machine learning (ML) and artificial intelligence (AI), has great potential to improve the performance of agricultural decision support systems. With this integration, we can achieve a higher level of automation, efficiency, and accuracy in analyzing remote sensing data. This will result in better allocation of resources and more targeted interventions. Numerous areas require significant improvements or modifications in current remote sensing systems, as well as techniques and methods for obtaining valuable insights from remote sensing images. Modern satellites re□uire an upgrade in sensor specifications to accurately estimate crop yield at larger scales, contributing to the achievement of sustainable development goals such as overall sustainability; good health and well-being; zero hunger; and poverty reduction. Forecasting crop yields and climate has been a significant challenge in developing and underdeveloped regions, which reduces the ability of societies to withstand impending disasters. High-altitude areas often experience frequent atmospheric disturbances, which can hinder the effectiveness of agricultural studies. As a result, remote sensing systems require more secure algorithms to minimize the impact of atmospheric distortions. In addition, it is important to note that in order to achieve the United Nations Sustainable Development Goals, it is crucial to have access to remote sensing data. Unfortunately, this data is still not widely available to most end users, particularly high-resolution data. It is necessary to implement mechanisms for improved and convenient access to satellite data.

REFERENCES

1. Alisaac, E., Behmann, J., Kuska, M. T., Dehne, H. W., & Mahlein, A. K. (2018). Hyperspectral uantification of wheat resistance to *Fusarium* head blight: Comparison of two *Fusarium* species. *European Journal of Plant Pathology, 152,* 869–884.
2. Azam, A., Ahmed, A., Wang, H., Wang, Y., & Zhang, Z. (2021). Knowledge structure and research progress in wind power generation (WPG) from 2005 to 2020 using CiteSpace-based scientometric analysis. *Journal of Cleaner Production, 295,* 126496.
3. Ban, T., Kawad, N., Yanagisawa, G., & Takezaki, A. (2008). Progress and prospects of resistance breeding to *Fusarium* head blight in Japan. *Cereal Research Communications, 36*(Supplement-6), 23–29.
4. Barbedo, J. G. A., Tibola, C. S., & Lima, M. I. P. (2017). Deoxynivalenol screening in wheat kernels using hyperspectral imaging. *Biosystems Engineering, 155,* 24–32.
5. Barbedo, J. G., Tibola, C. S., & Fernandes, J. M. (2015). Detecting *Fusarium* head blight in wheat kernels using hyperspectral imaging. *Biosystems Engineering, 131,* 65–76.
6. Behrend, J., & Eulerich, M. (2019). The evolution of internal audit research: A bibliometric analysis of published documents (1926–2016). *Accounting History Review, 29*(1), 103–139.
7. Chang, Y. W., Huang, M. H., & Lin, C. W. (2015). Evolution of research subjects in library and information science based on keyword, bibliographical coupling, and co-citation analyses. *Scientometrics, 105,* 2071–2087.
8. Chen, X., & Liu, Y. (2020). Visualization analysis of high-speed railway research based on CiteSpace. *Transport Policy, 85,* 1–17.
9. Delwiche, S. R., Rodriguez, I. T., Rausch, S. R., & Graybosch, R. A. (2019). Estimating percentages of *Fusarium*-damaged kernels in hard wheat by near-infrared hyperspectral imaging. *Journal of Cereal Science, 87,* 18–24.
10. Dong, F., Xu, J., Zhang, X., Wang, S., Xing, Y., Mokoena, M. P., & Shi, J. (2020). Gramineous weeds near paddy fields are alternative hosts for the *Fusarium graminearum* species complex that causes *Fusarium* head blight in rice. *Plant Pathology, 69*(3), 433–441.
11. Femenias, A., Bainotti, M. B., Gatius, F., Ramos, A. J., & Marín, S. (2021). Standardization of near-infrared hyperspectral imaging for wheat single kernel sorting according to deoxynivalenol level. *Food Research International, 139,* 109925.

12. Feng, Z. H., Wang, L. Y., Yang, Z. Q., Zhang, Y. Y., Li, X., Song, L., & Feng, W. (2022). Hyperspectral monitoring of powdery mildew disease severity in wheat based on machine learning. *Frontiers in Plant Science, 13,* 828454.
13. Guo, W., Yang, Y., Zhao, H., Song, R., Dong, P., Jin, Q., & Yang, Z. (2021). Winter wheat take-all disease index estimation model based on hyperspectral data. *Applied Sciences, 11*(19), 9230.
14. Haile, J. K., N'Diaye, A., Walkowiak, S., Nilsen, K. T., Clarke, J. M., Kutcher, H. R., & Pozniak, C. J. (2019). *Fusarium* head blight in durum wheat: Recent status, breeding directions, and future research prospects. *Phytopathology, 109*(10), 1664–1675.
15. Huang, L., Li, T., Ding, C., Zhao, J., Zhang, D., & Yang, G. (2020). Diagnosis of the severity of *Fusarium* head blight of wheat ears on the basis of image and spectral feature fusion. *Sensors, 20*(10), 2887.
16. Huang, L., Wu, Z., Huang, W., Ma, H., & Zhao, J. (2019). Identification of *Fusarium* head blight in winter wheat ears based on Fisher's linear discriminant analysis and a support vector machine. *Applied Sciences, 9*(18), 3894.
17. Jaillais, B., Roumet, P., Pinson-Gadais, L., & Bertrand, D. (2015). Detection of *Fusarium* head blight contamination in wheat kernels by multivariate imaging. *Food Control, 54,* 250–258.
18. Jin, X., Jie, L., Wang, S., Qi, H. J., & Li, S. W. (2018). Classifying wheat hyperspectral pixels of healthy heads and *Fusarium* head blight disease using a deep neural network in the wild field. *Remote Sensing, 10*(3), 395.
19. Liao, H., Tang, M., Luo, L., Li, C., Chiclana, F., & Zeng, X. J. (2018). A bibliometric analysis and visualization of medical big data research. *Sustainability, 10*(1), 166.
20. Liu, W. C., Liu, Z. D., Huang, C., Lu, M. H., Liu, J., & Yang, Q. P. (2016). Statistics and analysis of crop yield losses caused by main diseases and insect pests in recent 10 years. *Plant Prot, 42*(05), 1–9.
21. Liu, Z., Yin, Y., Liu, W., & Dunford, M. (2015). Visualizing the intellectual structure and evolution of innovation systems research: A bibliometric analysis. *Scientometrics, 103,* 135–158.
22. Ma, Z., Xie, Q., Li, G., Jia, H., Zhou, J., Kong, Z., & Yuan, Y. (2020). Germplasms, genetics and genomics for better control of disastrous wheat *Fusarium* head blight. *Theoretical and Applied Genetics, 133,* 1541–1568.
23. Mahlein, A. K. (2016). Present and future trends in plant disease detection. *Plant Dis, 100*(2), 1–11.

24. Mahlein, A. K., Alisaac, E., Al Masri, A., Behmann, J., Dehne, H. W., & Oerke, E. C. (2019). Comparison and combination of thermal, fluorescence, and hyperspectral imaging for monitoring *Fusarium* head blight of wheat on spikelet scale. *Sensors, 19*(10), 2281.
25. Mahlein, A. K., Steiner, U., Hillnhütter, C., Dehne, H. W., & Oerke, E. C. (2012). Hyperspectral imaging for small-scale analysis of symptoms caused by different sugar beet diseases. *Plant Methods, 8,* 1–13.
26. McMullen, M., Bergstrom, G., De Wolf, E., Dill-Macky, R., Hershman, D., Shaner, G., & Van Sanford, D. (2012). A unified effort to fight an enemy of wheat and barley: *Fusarium* head blight. *Plant Disease, 96*(12), 1712–1728.
27. Nielsen, L. K., Cook, D. J., Edwards, S. G., & Ray, R. V. (2014). The prevalence and impact of *Fusarium* head blight pathogens and mycotoxins on malting barley quality in the UK. *International Journal of Food Microbiology, 179,* 38–49.
28. Parry, D. W., Jenkinson, P., & McLeod, L. (1995). *Fusarium* ear blight (scab) in small grain cereals—a review. *Plant Pathology, 44*(2), 207–238.
29. Qian, J., Law, R., & Wei, J. (2019). Knowledge mapping in travel website studies: A scientometric review. *Scandinavian Journal of Hospitality and Tourism, 19*(2), 192–209.
30. Ropelewska, E., & Zapotoczny, P. (2018). Classification of *Fusarium*-infected and healthy wheat kernels based on features from hyperspectral images and flatbed scanner images: A comparative analysis. *European Food Research and Technology, 244,* 1453–1462.
31. Shah, D. A., De Wolf, E. D., Paul, P. A., & Madden, L. V. (2019). Functional data analysis of weather variables linked to *Fusarium* head blight epidemics in the United States. *Phytopathology, 109*(1), 96–110.
32. Singh, C. B., Jayas, D. S., Paliwal, J., & White, N. D. G. (2007). Fungal detection in wheat using near-infrared hyperspectral imaging. *Transactions of the ASABE, 50*(6), 2171–2176.
33. Spanic, V., Lemmens, M., & Drezner, G. (2010). Morphological and molecular identification of *Fusarium* species associated with head blight on wheat in East Croatia. *European Journal of Plant Pathology, 128,* 511–516.
34. Torres, A. M., Palacios, S. A., Yerkovich, N., Palazzini, J. M., Battilani, P., Leslie, J. F., & Chulze, S. N. (2019). *Fusarium* head blight and mycotoxins in wheat: Prevention and control strategies across the food chain. *World Mycotoxin Journal, 12*(4), 333–355.

35. Vrešak, M., Halkjaer Olesen, M., Gislum, R., Bavec, F., & Ravn Jørgensen, J. (2016). The use of image-spectroscopy technology as a diagnostic method for seed health testing and variety identification. *PloS One, 11*(3), e0152011.

36. Whetton, R. L., Hassall, K. L., Waine, T. W., & Mouazen, A. M. (2018). Hyperspectral measurements of yellow rust and *Fusarium* head blight in cereal crops: Part 1: Laboratory study. *Biosystems Engineering, 166,* 101–115.

37. Whetton, R. L., Waine, T. W., & Mouazen, A. M. (2018). Hyperspectral measurements of yellow rust and *Fusarium* head blight in cereal crops: Part 2: On-line field measurement. *Biosystems Engineering, 167,* 144–158.

38. Xu, X. M., Nicholson, P., Thomsett, M. A., Simpson, D., Cooke, B. M., Doohan, F. M., & Edwards, S. G. (2008). Relationship between the fungal complex causing *Fusarium* head blight of wheat and environmental conditions. *Phytopathology, 98*(1), 69–78.

39. Xue, A. G., Chen, Y., Seifert, K., Guo, W., Blackwell, B. A., Harris, L. J., & Overy, D. P. (2019). Prevalence of *Fusarium* species causing head blight of spring wheat, barley, and oat in Ontario during 2001–2017. *Canadian Journal of Plant Pathology, 41*(3), 392–402.

40. Zhang, N., Pan, Y., Feng, H., Zhao, X., Yang, X., Ding, C., & Yang, G. (2019). Development of *Fusarium* head blight classification index using hyperspectral microscopy images of winter wheat spikelets. *Biosystems Engineering, 186,* 83–99.

41. Zheng, Q., Huang, W., Cui, X., Dong, Y., Shi, Y., Ma, H., & Liu, L. (2018). Identification of wheat yellow rust using optimal three-band spectral indices in different growth stages. *Sensors, 19*(1), 35.

Chapter 2

INTRODUCTION TO CROP MANAGEMENT

Contents

2.1. INTRODUCTION

Over the past decade, there have been significant changes in agricultural production strategies. Such changes are being influenced by economic factors, aiming to minimize costs and maximize profits, as well as environmental regulations that promote the efficient and safe usage of agricultural chemicals. However, growers now have a heightened sensitivity to concerns over the quality, nutritional value, and safety of agricultural products. They carefully choose cultivars and make adjustments to planting dates based on expected weather patterns, such as La Niña or El Niño events. They have used biotechnological innovations as well to control pests, such as Roundup and insect-protected (Bt) ready crops. Moreover, the potential to sell carbon credits to various industries is revitalizing conservation tillage practices on-farm which improves carbon sequestration (Adams et al., 1993).

One of the most visible developments in the agricultural sector over the past decade has been the adoption of site-specific or precision crop management techniques (National Research Council, 1997). Growers have long recognized within-field variability in potential productivity. Now, at the beginning of the 21st Century, they are seeking new ways to exploit that variability. In the process, they are discovering they need more information on soil and plant conditions than was required a decade ago. Not only does this information need to be accurate and consistent across their farm and from year to year, but it must also be available at temporal and spatial scales that match rapidly evolving capabilities to vary cultural procedures, irrigations, and agrochemical inputs (Adams et al., 2000).

Extensive research conducted over nearly 40 years has consistently shown that a significant amount of the necessary information can be accessed remotely through satellite and aircraft-based sensor systems. With the incredible advancements in technology such as GPS receivers, microcomputers, RS (RS) technology, crop simulation models, GIS, and yield monitors has the power to revolutionize how growers run their lands and use precision farming methods (Adamsen et al., 1999).

2.2. BIOPHYSICAL BASIS FOR AGRICULTURAL RS

The contemporary applications of agricultural RS can be traced back to the groundbreaking work of ARS scientists Joseph Woolley, Harold Gausman, and William Allen. Their research laid the foundation for understanding how the optical properties of crop plants are related to their morphological characteristics. The scientists and their groups have extensively issued Hi-Res spectral signatures for various cultivated as well as natural breeds. These signatures help confirm spectral features related to normal growth conditions in plants, as well as such due to nutrient deficiency, and abiotic stresses along pests. Some notable publications include (Gausman, 1974).

2.2.1. Spectral Reflectance Properties of Leaves

The lant leaves usually have a low transmission and reflectance in the visible spectrum from 400–700 nm because of the absorption by accessory and photosynthetic plant pigments. Conversely, the transmission and reflectance levels are typically higher within near-infrared regions (NIR) from 700–1300 nm. This is due to the minimal absorptance by pigments or subcellular particles and the significant scattering at the mesophyll cell wall interfaces. The stark difference in how NIR and visible wavelengths reflect light is crucial for various remote methods used to monitor and manage natural vegetation and crop communities (Allen et al., 1998).

During periods of plant stress or at the end of the growing season, chlorophyll concentrations tend to decrease, which allows other leaf pigments like carotenes and xanthophylls to become more prominent. This phenomenon results in a widening of the green reflectance peak, which is typically found around 550 nm, towards longer wavelengths. This leads to a noticeable increase in visible reflectance, as observed by Adams et al. (1993), and gives the tissues a chlorotic appearance. Simultaneously, the NIR reflectance decreases, although to a lesser extent than the visible increases. As stress levels rise, there is a noticeable shift in the transition between NIR and visible green vegetation, known as the "red edge." This shift moves towards shorter wavelengths and can even disappear completely in senescent vegetation.

The optical properties of the leaves in the middle of the shortwave-infrared (SWIR) between 1300 and 2500 nm section of the solar spectrum are significantly affected by the presence of water in the tissues. Reflectance within this section is typically higher for vigorously growing vegetation, however, it tends to decrease as the plants lose moisture. Research, however, indicates that these drought-induced reductions in SWIR reflectance may not be large enough to offset physiologically meaningful modifications to plant water content for this wavelength interval to be used effectively in the field for a diagnosis of water stress (Bauer, 1975).

2.2.2. Spectral Reflectance Properties of Soils

When it comes to plants, the spectral signatures in agricultural soils tend to be quite straightforward. Across the visible and near-infrared regions, their reflectance typically increases monotonically. Dry and smooth-surfaced soils are generally brighter, while soils with high levels of organic matter and water generally have lower reflectances. Certain minerals found in soil can exhibit distinct spectral characteristics, such as increased red reflectance when iron oxides are present. Within the SWIR range, soil spectra exhibit a greater number of features compared to shorter wavelengths. However, these features are still primarily influenced by the mineral and water content and litter. When crop residue is present, it leads to noticeable alterations in the reflectance in comparison with the bare soils and the partial plant canopies. Hence, it is crucial to consider the presence of residue when making observations across various soil types and crop production methods. This has been highlighted in studies conducted by Nagler et al. (2000); and Daughtry et al. (2001). An in-depth review by Barnes et al. (2003) explores the utilization of different RS techniques in soil management, specifically in defining zones for crop management.

2.2.3. Crop Canopies and VI

It is expected that the spectral signatures of the field crop canopies can be intricate and may differ significantly from that of individual green leaves checked under controlled light conditions (Figure 2.1). Even as the season progresses, the canopy spectra undergo dynamic changes due to variations in the proportions of vegetation and soil, as well as the plant components' varying arrangements. Vegetation Indices (Vis) are an elegant and simple way to extract the signal for the number of green plants from the complex canopy spectra. Vis uses the basic differences between plant and soil spectra that were discussed earlier. They are often calculated as linear combinations, differences, or ratios of reflected light in the NIR and visible bands. The indices like the normalized difference vegetation index given by NDVI = (NIR – Red)/(NIR + Red) and the ratio vegetation index is given by RVI = NIR/Redwork—incredibly well when management objective calls for a quantitative way to track leaf or green biomass area index over the season or identify irregular growth patterns within fields. Specialized indices like SAVI and modified SAVI have been established to reduce the impact of different background soil reflectance properties on the accuracy of VI (Bausch & Diker, 2001).

Since VI is positively correlated with both the leaf area index and green biomass of crop canopies, they have been the foundation for numerous RS applications to crop management (Figure 2.3(a)). Vis has also been demonstrated to provide reliable estimates of the fractional amount of net radiation entering soil heat flux. Vis are often employed interchangeably with plant vigor or health. This may be misleading since broad waveband Vis usually does not have the diagnostic power to identify a specific type of stress or explain why biomass is at a given level. Reflectance indices such as the Water Band Index (WBI),

Photochemical Reflectance Index (PRI), and Normalized Pigment Chlorophyll Ratio Index (NPCI) are narrower band indices that show correlation with specific physiological plant responses. These indices hold potential for diagnosing nutrient and water stress, as suggested by studies conducted by Peñuelas et al. (1994) and Gamon et al. (1997). Using a specialized index called the canopy chlorophyll content index or CCCI, experts in crop management can accurately identify nutrient stress in cotton by analyzing a combination of a vegetation index and reflectance in a rather narrow red edge band (i.e., ~720 nm). This method helps differentiate nutrient stress from other factors that may lead to a decrease in green biomass.

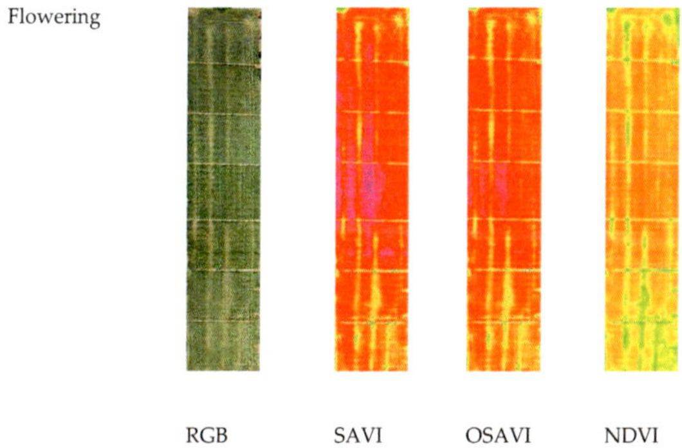

Figure 2.1. *RGB, soil-adjusted vegetation index (SAVI), optimized soil-adjusted vegetation index (OSAVI), and normalized difference vegetation index (NDVI) maps at different crop stages: Seedling, tillering, and flowering growth stages.*

Source: Marino, S., & Alvino, A. (2021). Vegetation indices data clustering for dynamic monitoring and classification of wheat yield crop traits. Remote Sensing, 13(4), 541. https://doi.org/10.3390/rs13040541 (Creative Commons License).

There have been different approaches, such as hyperspectral analysis, which have been suggested and experimented with to identify stress in plants caused by water, nutrients, and pests. These methods aim to minimize interference from soil conditions or biomass levels. These methods typically utilize peak fitting procedures, derivative analysis, and ratio analysis to link spectral features with specific stressors. When it comes to understanding the connections between hyperspectral data and plant properties, advanced statistical methods like partial least-squares regression analysis, principal component analysis, neural networks, and fuzzy logic can be utilized break down images to their individual components (e.g., healthy and stressed plant areas, shaded, and sunlit soil) by using a library of "pure" hyperspectral signatures of scene components, also called endmembers (Arslan & Colvin, 2002).

2.2.4. Emitted Thermal Radiation

Every object on the infra-surface of Earth emits radiation in the thermal-infrared region (TIR) of the spectrum, which falls within the range of approximately 8–14 mm. Understanding the emitted energy from an object can be incredibly valuable in evaluating the crop water stress. This is because the surface temperature in plants is strongly moderated by the soil water availability and its impact on crop evapotranspiration (Barnes et al., 2003).

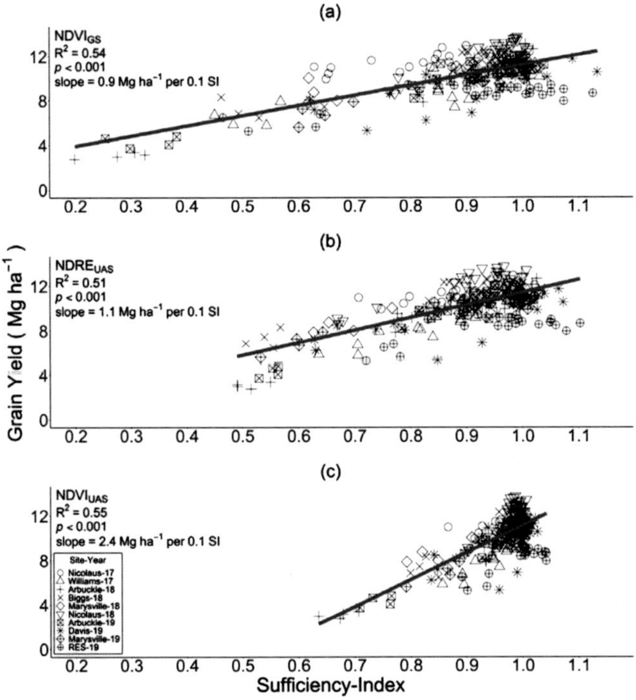

Figure 2.2. *The relationship between (a) GreenSeeker normalized difference vegetation index ($NDVI_{GS}$) sufficiency-index (SI); (□) unmanned aircraft system (UAS) normalized difference red-edge index ($NDRE_{UAS}$) SI; and (c) $NDVI_{UAS}$ SI measured at panicle initiation (PI) rice growth stage and grain yield as described by linear mixed-effects models. The coefficient of determination (R^2) reported in each panel represents the proportion of varia□ility explained □y the model fixed effects only.*

Source: Rehman, T. H., Lundy, M. E., & Linquist, B. A. (2022). Comparative sensitivity of vegetation indices measured via proximal and aerial sensors for assessing N status and predicting grain yield in rice cropping systems. Remote Sensing, 14(12), 2770. https://doi.org/10.3390/rs14122770 (Creative Commons License).

After noting the significant differences between plant temperatures and air temperature, researchers from ARS explored the environmental factors that influence crop temperature and started considering its potential as an indicator of water stress. With the introduction of affordable and easily accessible infrared thermometers in the mid-70s, scientists from the ARS swiftly embraced this new technology. They transitioned from the thermocouples to measure the plant temperatures to a range of non-contact methods to evaluate water status along with forecast crop yields across larger areas (Bartholic et al., 1972) (Figure 2.2).

Various descriptive terms were created to characterize the thermal indices employed in these methods. The agronomic literature first reported the following measures of plant stress caused by water stress: "Water Deficit Index" (WDI); "Thermal Kinetic Window" (TKW); "Non-water-stressed baselines" (Idso, 1982); "Crop Water Stress Index" CWSI; (Jackson et al., 1982); "Stress-Degree-Day" (SDD). Research has indicated that various factors, such as physical and biological stresses, can impact transpiration and lead to higher plant temperatures. These stresses are often linked to the water status of the plants and can result in reduced potential yield. The TIR is a crucial factor in the surface energy balance and thus is widely utilized in remote methods for evaluating evapotranspiration. Various studies have extensively explored this area (Maas, 1988) (Figure 2.3).

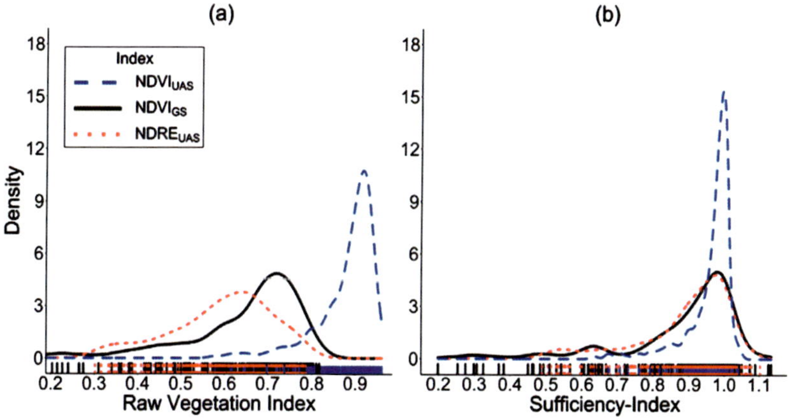

Figure 2.3. *Kernel density distributions of unmanned aircraft system (UAS) normalized difference vegetation index ($NDVI_{UAS}$), GreenSeeker (GS) NDVI ($NDVI_{GS}$), and UAS normalized difference red-edge index ($NDRE_{UAS}$) – (a) raw vegetation index (VI); and (□) sufficiency-index (SI) measured at the panicle initiation (PI) rice growth stage.*

Source: Rehman, T. H., Lundy, M. E., & Linquist, B. A. (2022). Comparative sensitivity of vegetation indices measured via proximal and aerial sensors for assessing N status and predicting grain yield in rice cropping systems. Remote Sensing, 14(12), 2770. https://doi.org/10.3390/rs14122770 (Creative Commons License).

2.2.5. Exogenous Factors Affecting Remote Observations

It is crucial to understand that evaluating crop growth and plant response to environmental stress remotely is far more complicated and involved than identifying chemicals in vitro using their spectral absorption characteristics. Due to the age of the individual tissues and the organs' architectural arrangement, the thermal and optical properties of the plant canopies vary with the growth stage (Figure 2.1). Additionally, factors that are indirectly related to either the biophysical or the agronomic plant properties, such as row orientation, meteorological phenomena, topography, viewing angles, and illumination have a significant impact. One big problem with using RS for farming is being able to tell the difference between spectral signals that come from a plant responding to a certain stress and signals that come from "background noise" or normal plant biomass that comes from exogenous non-plant factors. General relationships between plant response and spectral properties can be achieved, according to results from several crops grown in various locations (Masoni et al., 1996).

2.3. WATER MANAGEMENT

Substandard irrigation timing plus inadequate water applications are prevalent elements that restrict production in numerous semi-arid and even arid agricultural regions. Consequently, a significant amount of ARS research has concentrated on RS techniques that monitor plant water status, measure evapotranspiration rates, and estimate crop coefficients to determine the extent and periods to irrigate (Bowman, 1989).

2.3.1. Plant Water Status

Scientists at the ARS have developed, tested, and refined a variety of non-invasive thermal indices to assess if plants are transpiring enough water from the atmosphere and to deduce the water status of the plants based on that measurement. According to Gardner et al. (1992), these indices, which comprise the aforementioned CWSI, SDD, and WDI, have been applied to studies involving over 40 distinct crop species. Even though each index varies in how complicated it is and how many extra crop-specific and meteorological parameters are needed, they all depend on the plant temperatures that can be gathered from afar using thermal imaging devices or radiation thermometers (Gardner et al., 1992; Millard et al., 1990).

The fundamental ideas are straightforward. As plants deplete the water in the soil, transpiration slows down. This means that plants get warmer compared to the air temperature or the temperature of a reference crop that has been well watered. The entire range of plant temperature variations caused by soil water availability depends on the atmosphere's evaporative demand as well as the transpiration characteristics unique to each crop. As suggested by Idso et al. (1982), lower and upper boundary temperatures, also called "baselines," may be determined empirically from the previous field observations of well-watered and stressed canopies, or they can be estimated using theoretical energy balance calculations according to Jackson et al. (1982). Additionally, they may originate from a blend of theoretical and/or empirical methods. The majority of research has demonstrated that reflectance in the visible, near-infrared, or short-wavelength infrared regions is less sensitive to acute water stress than is the thermal infrared. But the Vis and reflective spectrum also react to plant water status as it causes a shift in the canopy architecture, such as wilting or rolling of the leaves, or when there is persistent water stress that causes growth to slow down, decreases the green leaf area index (GLAI), or modifies the rates of senescence (Carlson et al., 1995). In areas where extra water is required to grow a crop, thermal plant water stress indices usually give enough notice to schedule irrigations. However, a sufficient amount of evaporative demand from the atmosphere, a sufficient amount of water holding capacity in the soil, plus irrigation depth are necessary for the technique to be applied successfully. The TIR is not as useful for planning irrigation in mesic areas because there is less evaporative demand there, so there aren't as many temperature differences between stressed and well-watered plants. In these circumstances, the water stress signal may be obscured by measurement errors and variations in plant temperatures brought on by variations in wind speed. However, even in humid areas, thermal methods can help determine crop stress in certain areas of the field due to spatial soil variation or prolonged dry spells. Thermal indices were found to be less effective by Benasher et al. (1992) in the management of micro-irrigation drip systems, in which the quantity of soil water replenished during every irrigation was comparatively small with the crop's daily requirements.

Thermal indices may overestimate the water stress when sensors observe a combination of warm soil temperatures and cool plant temperatures due to incomplete canopy cover. This issue can be reduced for ground-based measurements by limiting observations to the elements of transpiring foliage or by adopting an oblique observation angle. However, in nadir data from the overhead sensors, mixed pixels are frequently unavoidable. A clever solution to this issue is the Water Deficit Index, which combines a VI (to account for the degree of plant cover) with the TIR. This method makes it easier to detect water stress early in the season, which helps with the irrigation schedules, and it also makes TIR from planes and satellites more useful (Burke, 1994). Literature regarding the agricultural RS is filled with examples of the application of thermal indices to plan irrigations in multiple crops, for example:

Crop	References
Bermuda grass	Jalalifarahani et al. (1993, 1994)
Corn	Fiscus et al. (1991); Nielsen and Gardner (1987); Wanjura and Upchurch (2000, 2002); Yazar et al. (1999).
Sorghum	Hatfield (1983a)
Alfalfa	Hutmacher et al. (1991); Moran et al. (1994)
Alover	Oliva et al. (1994a, b)
Cotton	Pinter and Reginato (1982); Reginato and Howe (1985); Shanahan and Nielsen (1987); Wanjura and Upchurch (2000, 2002)
Soybeans	Nielsen (1990)
Sunflowers	Nielsen (1994)

The majority of all thermal irrigation scheduling algorithms were created and then tested on field plots with infrared radiometers that are attached to the ground. Currently, satellites can only send thermal data through sensor systems that have a limited spatial resolution to be useful in irrigated agriculture. For example, Landsat 7's ETM+ has a 10.4–12.5-mm sensor boasting a spatial resolution of 60-m. Despite early indications of its potential utility, aircraft TIR has not been made available widely.

This is regrettable because the reflective portion of the spectrum does not provide the distinctive details on plant water status that the thermal infrared does. According to a cost/benefit analysis by Moran (1994), if the growers of an irrigation district join up and purchase imagery, the thermal infrared sensors on an aircraft for irrigation scheduling are feasible and reasonably priced.

Precision agriculture applications can benefit from techni□ues that use thermal imaging radiation (TRIR) to evaluate the spatial variation in soil water availability.

For instance, Hatfield et al. (1987) demonstrated how management practices affected surface temperature patterns in California's Central Valley and how such patterns were connected to uniformity in water applications. They discovered that the amount of water available affected the spatial variation of the surface temperatures in wheat and grain sorghum fields. They discovered that the variability of surface temperature increased as the soil water content fell below 50% of available, and they proposed that this might be employed as a possible management tool. The potential of using spatial variation as a water management tool has not yet been completely realized. An alternative would be to equip irrigation booms with infrared sensors so that the unit can monitor crop needs and modify irrigation volumes as it moves across the field (Carter, 1991).

2.3.2. Evapotranspiration (ET) and Crop Coefficients

ARS scientists have developed and tested methods at field, farm, and regional scales for evaluating the spatial and temporal dynamics of ET. Typically, these methods integrate remote measurements of radiation, both emitted and reflected, with ground-based meteorological observations. The residual in the energy balance equation is used to estimate latent energy (LE) exchange. On clear days, it is possible to convert the remote and almost instantaneous estimates of LE that were obtained around midday using information from satellite/aircraft into daily values with a fair amount of precision (Ehrler, 1973). The lack of thermal sensors on aircraft or satellite platforms with an appropriate spatial and temporal resolution has hindered the applications of these techniques, which have great potential for estimating water use over large areas (Hutmacher et al., 1991).

A different approach to monitoring plant water requirements uses multispectral proxies for Kcb in conjunction with regular meteorological estimates of potential evapotranspiration. Advanced irrigation scheduling methods, like the FAO-56 approach, utilize Kcb values. These values represent the ratio of the actual crop evapotranspiration to the potential evapotranspiration of the reference crop, like grass or alfalfa that is grown under ideal agricultural conditions. Obtaining Kcb from published curves or tables can limit their flexibility in accounting for variations in crop water needs (These variations can be caused by factors such as unusual weather patterns, non-uniform water application, differences in plant population, pest pressures, or nutrient stresses. Scientists at ARS noticed the correlation of Kcb behavior with the seasonal trajectory of multispectral Vis. They suggested and later proved that these Vis can be used for irrigation scheduling. Several studies have been conducted on this topic. With the inclusion of VI surrogates for Kcb in scheduling programs, growers can receive valuable feedback from plants. This feedback allows them to make better adjustments to irrigation timings and extents, helping to prevent critical soil water deficits. Additionally, it opens up the potential for fine-tuning the precision irrigation system (Gardner et al., 1992).

2.3.3. Salinity Stress

Salts in irrigation water and soils play a critical role in determining the productivity of croplands, as highlighted by Rhoades et al. (1989). Mapping the affected areas in time and space is essential for implementing effective remedial solutions. This can be achieved through the use of RS measurements, which detect reduced biomass or the changes in the plants' spectral properties growing in affected areas, or they can identify the contaminated soils through unnaturally high surface reflectance factors. There are strong links between Vis in the middle of the growing season and the ultimate yields of sorghum and cotton crops, which are affected by salt stress on a sub-field level. Research has indicated that plants exposed to high levels of salt in irrigation water may experience an increase in canopy temperature. This finding suggests that it may be possible to detect stress in plants before it becomes visible and take action by raising the leaching fraction or using higher-quality water (Clarke, 1997).

2.3.4. Thermal Kinetic Window

Scientists from the ARS observed that transpirational cooling plays a crucial role in keeping the tissue temperatures of irrigated crops significantly below dangerous levels, specifically in arid regions where they are frequently subjected to sensible and intensely radiant heat. These findings were reported by Burke et al. (1994) and Hatfield et al. (1987) among others. They used the apparent Michaelis constant (Km) along with the variable fluorescence to find the "Thermal Kinetic Window" (TKW) range of temperatures where the biochemical processes in the tissues were working at their optimal rates. Also, they suggested that infrared thermometers may be employed to find out if plants were in the TKW range. The TKW varies depending on the species, which aligns with the crop's seasonal growth patterns and geographical distribution. For example, the TKW for wheat is approximately 20°C, whereas for cotton it is around 28°C (Clawson et al., 1989). The ARS team made an interesting discovery regarding the correlation between biomass production, final yields, and the duration of time a crop spent in its TKW. They further developed and patented a sensor system known as BIOTIC, which, using this concept effectively controls micro-irrigations and addresses certain issues mentioned by Benasher et al. (1992).

2.4. NUTRIENT MANAGEMENT

Efficiently managing nutrients is a significant challenge in the realm of production agriculture. With the help of RS, field-scale diagnostic methods are now available to detect nutrient deficiencies at an early stage, preventing any potential loss in yield or quality. When used in conjunction with variable rate sprayer equipment, canopy sensors can provide real-time data on specific application needs. This helps reduce contamination of groundwater and/or surface supplies also enhancing the efficiency of nutrient use (Corp et al., 1997).

2.4.1. Nitrogen

Having sufficient nitrogen (N) is crucial for successful crop production in today's agriculture. Unfortunately, nitrogen is frequently applied in excessive amounts without considering the specific needs of the crop or the potential environmental consequences, simply to ensure that there are sufficient levels available. For instance, let us consider corn cultivation in the upper Midwestern United States. It is important to time the application of nitrogen (N) to align with the crop's peak uptake. However, the practice of using leaf tissue testing to assess crop requirements is not commonly used in this region (McMurtrey et al., 1994). As a result, fields may be frequently over-fertilized. Various methods have been developed to assess spatial variation in nitrogen concentrations across corn fields. These include the use of color photography. SPAD chlorophyll meter, and canopy reflectance factors. These methods relied on comparing readings from a properly fertilized strip in the exact field. This meant that strenuous knowledge of the association between nutrient concentration and crop reflectance, as well as the need for precise sensor calibration and the conversion of data to surface reflectance factors, were not needed ahead of time (Osborne et al., 2002).

In the Great Plains, over 50% of the nitrogen that corn needs is usually added before it is planted. If one only applies minute amounts of fertilizer "as required" during the season, that can stop nitrogen from being washed away by rain or too much watering. This N reflectance index (NRI) was created by Bausch (2001) using NIR and green reflectance of a crop of irrigated corn. The NRI showed a strong correlation with an N sufficiency index derived from SPAD chlorophyll meter data. This allowed for a quick evaluation of the nitrogen status of corn plants, making it useful for mapping. According to a more recent study conducted by Bausch and Diker in 2001, the use of the NRI to track in-season plant N led to a significant reduction of applied N through fertigation. This reduction amounted to 39 kg N ha^{-1} without any negative impact on grain yield. With its focus on plant canopy rather than individual leaf measurements obtained through SPAD readings, this index holds promise for broader applications and can directly contribute to the advancement of variable rate fertilizer application technology (Nouvellon et al., 2001).

With a deep understanding of crop management, proposed a clever strategy. They suggested that by using a remote mid-season estimate of potential yield, growers could make more informed decisions about adjusting topdress N applications. This approach would take into account mineralization rates throughout the season. Pre-plant soil N tests, and projected N removal. They calculated the possible grain yields of the winter wheat (*Triticum aestivum* L.) based on multiple post-dormancy NDVI measurements. These measurements were adjusted by the number of growing degree days numbers which had accumulated in-between the dates of observation. With this normalization, adjustments were made to account for variations in local weather and to compensate for differences in management options and soil properties that impacted early season growth and stand establishment. Possibly, these approaches could be applied in any situation where it is possible to remotely predict crop yield (Schepers & Francis, 1998).

2.4.2. Other Nutrients

It can be challenging to monitor symptoms that are caused by other nutrient deficiencies, as they hardly present themselves uniformly across the fields and must frequently be differentiated from background variation in the canopy density. In their research, Osborne et al. (2002a, b) found that hyperspectral data can be valuable in identifying variations in nitrogen (N) and phosphorus (P) levels in both leaves and canopies. However, it is important to note that these relationships may not remain consistent throughout all plant growth stages.

Adams et al. (2000) have successfully identified deficiencies in soybean leaves by utilizing advanced techniques such as hyperspectral reflectance and leaf fluorescence. By analyzing the reflectance spectrum of 570–670 nm, they were able to evaluate leaf chlorosis and detect deficiencies in essential nutrients such as Fe, Mn, Zn, and Cu. Using advanced analysis tools and hyperspectral imaging sensors, such as spectral mixing techniques and partial least squares regression, researchers can now easily conduct studies at the canopy level (Beerwinkle et al., 1993).

It is worth noting that scientists from ARS have collaborated with various organizations, including NASA, EPA, and USGS, to enhance RS technologies for identifying alterations in plant biochemistry, physiology, and metabolism. For instance, they conducted initial research on using plant fluorescence to identify water stress in citrus.

Recent advancements in laser-induced fluorescence (LIF) techniques show promise in identifying nutrient and water stress, as well as determining the ideal levels of plant growth and yield based on varying fertilization rates in field conditions. These methods have been extensively studied and documented by researchers (Choudhury et al., 1994).

2.5. PEST MANAGEMENT

With RS, it becomes remarkably easy to identify unusual areas in an orchard or a field that have been impacted differently by weeds, pests, or diseases. Indeed, over 35 years ago, ARS scientists already utilized aerial color-infrared photography to achieve this objective. They would then analyze their findings by comparing them to lab spectra of leaves damaged by pests (Dupont et al., 2000).

2.5.1. Weeds

Weeds can be among significant management costs for growers as they fight with the crops for essential resources like light, nutrients, and water. It often leads to a decrease in crop yield and quality, resulting in higher management costs. Poorly timed or improper herbicide applications may result in unintended consequences for the environment and crop performance. Recently, the world has seen a noticeable shift in weed control methods. Instead of relying on uniform, early-season options, farmers are now opting for herbicide-ready crops along with the application of post-emergence herbicides just as necessary. Using RS to identify weed patches within fields has generated significant interest. This strategy allows for precise targeting of these areas using variable rate ground and aerial spray rigs. By implementing these strategies, one can effectively minimize the need for herbicides in areas free of weeds, thus reducing the risk of groundwater contamination while still maintaining effective weed control (Pinter et al., 2003).

Understanding the kind of weed is crucial when customizing herbicide options to determine treatment rates. Gausman et al. (1974) conducted preliminary laboratory studies that identified variations in the optical properties of weeds among different species. Subsequent research by Richardson et al. (1985) showed that uniform plots of pigweed and Johnsongrass could be distinguished from cantaloupe, sorghum, and cotton plots using multispectral aerial video images. Speculation suggests that with advancements in technology and the availability of more precise data, it may be possible to obtain real-time info on the weed infestations within the crop canopies. It has proven to be highly effective in the management of weed species like leaf spurge and salt cedar in grazing areas and wildlands, as highlighted in a review by Hunt et al. (2003). Dickson et al. (1994) pioneered a method for the identification of wild proso millet and velvetleaf weeds in corn fields. This method utilized neural networks, visible wavelengths digital images, and the weed patches' spatial characteristics. Their approach demonstrated an impressive accuracy of over 90% when evaluated on an independent data set (Hart & Myers, 1968).

In their study, effectively reduced herbicide usage and production costs by utilizing spectral contrasts between the bare soil and the green plants. By doing so, they were able to implement real-time herbicide spraying just on the plants present between the rows of soybeans, effectively inhibiting the weeds. This method proved to be just as effective as conventional continuous-spray methods but with the added benefit of reducing production costs and herbicide usage. Machine vision techniques have been utilized to identify weed seedlings by analyzing their leaf shapes. Additionally, these techniques have been employed to guide an automatic precision herbicide sprayer (Tian et al., 1999).

A grower can benefit greatly from being able to identify unintentional herbicide damage to their crop when it comes to insurance or legal matters. Through their research, Hickman et al. (1991) found that it was not only possible to remotely detect and map moderate herbicide damage in cotton, but also estimate the application amounts. They compared visual assessment with a wide-band handheld radiometer, NIR videography, and color-infrared photography approaches. Donald (2000) utilized video photography to measure the arrest of soybean and corn plants caused by herbicide damage. With the help of a lab-based multi-spectral fluorescence imaging system (MFIS), Kim et al. (2001) successfully identified changes in soybean leaf fluorescence following herbicide treatment. Sudduth & Hummel (1993) created a portable NIR spectrophotometer to enhance the efficiency of herbicide application. This device is used to estimate soil organic matter, which is crucial in determining the appropriate amount of herbicide to be applied. RS provides valuable insights into field-scale analysis of herbicide injury problems and helps determine the optimal rates of herbicide application (Hanks et al., 1998).

2.5.2. Arthropod and Nematode Pests

In a series of renowned studies, scientists (Hart & Meyers, 1968) utilized color-infrared (CIR) photography and accompanying hyperspectral reflectance data to detect citrus orchard trees that were affected by the brown soft-scale insects (*Coccus hesperidum*). Monitoring changes in infestation levels was possible due to the scale of insects that excreted honeydew. This substance provided an ideal environment for a sooty mold fungus that had low reflectance in both NIR and visible wavelength regions. As the season progressed, this fungus tended to accumulate, allowing for effective monitoring (Gausman, 1981). Various techniques, such as multispectral videography and CIR film, have been successfully employed to detect pests like brown soft scale and citrus blackfly in citrus crops, as well as whitefly infestations in cotton (Figure 2.2).

During a greenhouse study conducted by Blackmer (1996), wheat seedlings were intentionally infested with green bugs known as *Schizaphis graminum* Rondani or the aphids known as *Diuraphis noxia* Mordvilko to examine the impact of sucking insects on leaf reflectance. After 3 weeks, they conducted measurements on the reflectance properties of the individual leaves using an external integrating sphere (Gamon et al., 1997). When comparing healthy plants to infested plants, it was observed that the infested plant leaves had lesser concentrations of chlorophyll. Additionally, notable changes in reflectance spectra were noticed at specific wavelengths, particularly between 500 and 525; also 625 and 635; and 680 and 695 nm). These findings indicate that canopy spectra could be valuable in identifying outbreaks in real-field scenarios

(Chappelle et al., 1984). By analyzing hyperspectral imagery captured during NASA's Airborne Visible Infrared Imaging Spectrometer (AVIRIS) flights over the cotton fields in California, Fitzgerald et al. (in press) successfully assessed the magnitude and impact of the strawberry spider mite (*Tetranychus turkestani* U.N.) damage across various fields. They created a reference library of the spectral signatures from the leaves affected by mites, which develop a reddish pigmentation, and also from healthy leaves and different types of soil. Using spectral mixing analysis, the hyperspectral AVIRIS images of the fields were decomposed ("unmixed") into components linked to the end members, which included the signatures of mite-stressed and healthy crops. Using this type of geo-referenced imagery across large areas, it becomes possible to accurately identify areas within fields that are affected by mites. This allows for more targeted pest scouting and the application of pesticides at variable rates. Using specific strategies for pest control not only decreases the overall use of pesticides but also creates refugia for helpful insects. These beneficial insects can then repopulate treated areas, reducing the risk of secondary pest infestations (Clothier et al., 1986) (Figure 2.4).

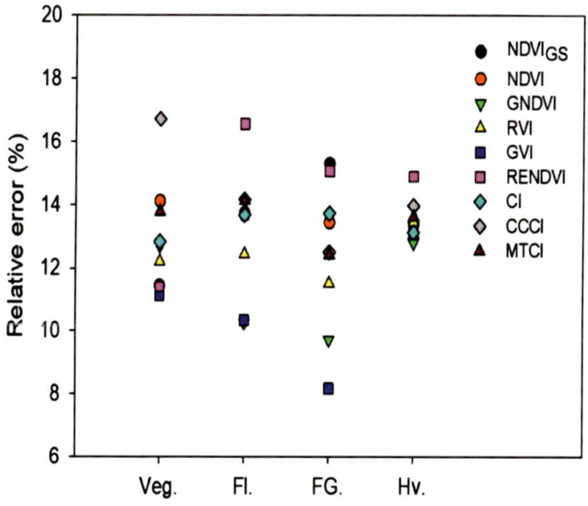

Figure 2.4. *Vegetation indices (Vis) can be useful tools to evaluate crop nitrogen (N) status.*

Source: de Souza, R., Peña-Fleitas, M. T., Thompson, R. B., Gallardo, M., & Padilla, F. M. (2020). Assessing performance of vegetation indices to estimate nitrogen nutrition index in pepper. Remote Sensing, 12(5), 763. https://doi.org/10.3390/rs12050763 (Creative Commons License).

Using a straightforward method, one can detect pests by observing the alterations in the biomass of green plants or GLAI resulting from root pruning, leaf skeletonizing, or herbivory. In aerial photographs or photos made from multispectral Vis, the pest problems show up like anomalous spots in the middle of healthy-looking plants. This approach tends to be more effective in monocultural field crops compared to natural ecosystems or a mixed crop. In a study conducted by ARS researchers, they used CIR film to assess the impact of soil fumigation and crop rotation on a nematode called *Rotylenchulus reniformis*, which is commonly found in cotton fields in Texas. In a study conducted by Cook et al. (1999), multitemporal NIR videography was employed to track the seasonal development of the southern root-knot nematode (*Meloidogyne incognita* Chitwood), and the soil-borne fungi complex it is associated with in Kenaf (*Hibiscus cannabinus* L.). It should be considered that there could be various elements affecting the reduced plant vigor, which may not necessarily be related to pests. Therefore, it is important to have further information like temporal, spatial, and spectral clues within a decision support system to accurately identify the problem.

Considering the current state of RS technologies, it seems unpromising that we will see the development of methods able to detect a tiny amount of important nematode or arthropod pests in the near future. Having a good understanding of when where and when they can be found can be □uite beneficial for guiding field scouts and implementing pre-emptive control measures. Various studies have utilized active radar systems to track the movement and flight patterns of insects that have significant economic impact, such as grasshoppers, noctuid moths, and honeybees. These studies have been conducted by experts in the field. Such information can be easily secured through the existing network of weather radars and can be used to alarm growers about the heightened risk to their local crops (Franz et al., 1991).

Large-scale aerial photography can be utilized to classify landscape features besides their correlation with loads of pests and their predators, as demonstrated by Elliott et al. (1999) in their study on the cereal aphid in South Dakota. Assuming that specific insects, such as the tarnished plant bug (*Lygus lineolaris*), had a higher tendency to feed on the fastest-growing parts of cotton fields, Willers et al. (1999) utilized NDVI images of commercial cotton fields. These images were used to assess the vigor of the crops and subsequently direct field scouts to those specific areas for targeted sampling. Using the scouting reports and imagery, plausible maps of insect abundance were constructed in a GIS. By utilizing GPS technology, the ground sprayer's controller was able to apply pesticide precisely to areas that posed a high risk. Through commercial field trials, this approach successfully decreased pesticide usage by ~40% and mitigated the general environmental toxic chemicals impact (Dupont et al., 2000) (Figure 2.5).

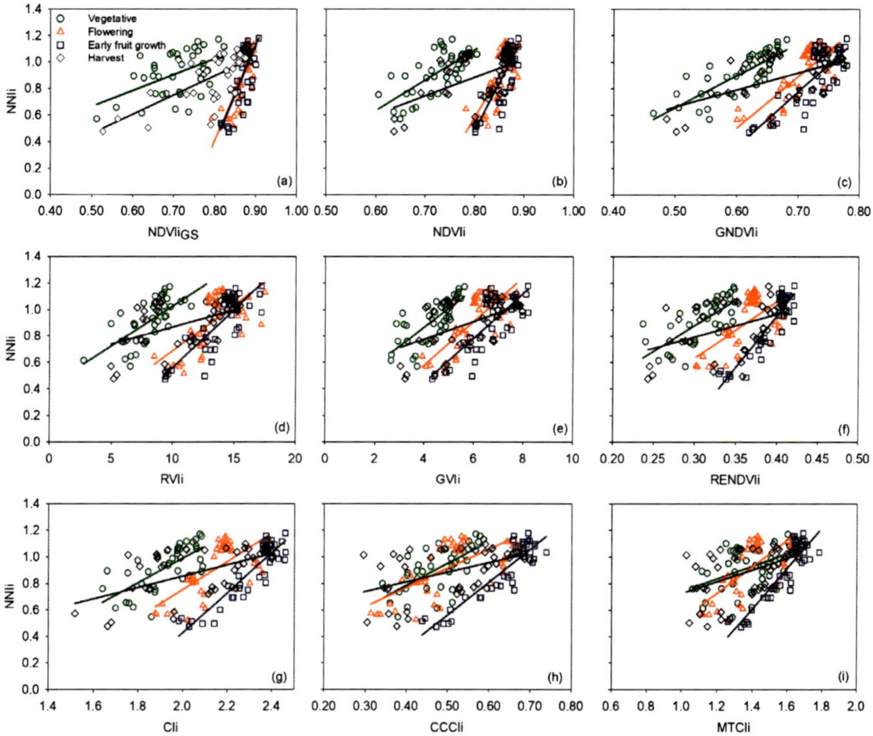

Figure 2.5. *Linear regressions between each integrated vegetation index and integrated nitrogen nutrition index (NNIi) for the four vegetative stages, for calibration data (n = 40). Circle: Vegetative; Triangle: Flowering; Square: Early fruit growth; and Diamond: Harvest stage. Panel (a) shows the normalized index vegetation index (NDVI) measured with the GreenSeeker sensor and the other panels (b–i) show indices calculated with the crop circle sensor.*

Source: de Souza, R., Peña-Fleitas, M. T., Thompson, R. B., Gallardo, M., & Padilla, F. M. (2020). Assessing performance of vegetation indices to estimate nitrogen nutrition index in pepper. Remote Sensing, 12(5), 763. https://doi.org/10.3390/rs12050763.

2.5.3. Disease

Examples of how ARS used RS technology to identify crop diseases and evaluate their impacts on productivity include identifying circular areas affected by *Phymatotrichum omnivorum*, or cotton root rot, and estimating yield losses in sugar beets due to black root disease. In a study conducted by Cook et al. (1999), it was shown that aerial video imagery can detect *P. omnivorum* in kenaf. This is particularly useful because kenaf has a tall growth habit that makes it extremely challenging to survey from the ground.

The TIR has the potential to detect diseases that obstruct the flow of water from the soil throughout the plant to the atmosphere at an early stage, potentially before visual inspection. For instance, in a study conducted by Pinter et al. (1986), it was observed that cotton plants with root infections caused by the soil-borne fungus *P. omnivorum* and sugar beets infected with *Pythium apanidermatum* exhibited sunlit leaf temperatures that were 3 to 5°C higher compared to nearby healthy plants. Researchers found the TIR to be valuable in identifying root disease in red clover when it was grown with irrigation. Extensive research is necessary when utilizing RS to identify particular diseases or differentiate them from other causes of plant stress (Fiscus et al., 1991). Utilizing hyperspectral techniques may offer some assistance, however, combining these techniques with weather-driven computer models of disease development is likely to provide the most effective approach (Peynado et al., 1980).

2.6. YIELD PREDICTION

Understanding yield is crucial as it reflects the combined impact of weather conditions and management strategies throughout the entire growing season. With RS approaches, growers can obtain accurate yield assessments and observe variations across their fields. They share similarities with combine-mounted yield monitors, which play a crucial role in precision agriculture (Feldhake & Edwards, 1992). Remote measurements have the advantage of being taken frequently during the season, giving valuable insights into growth rates and how plants respond to changing management practices and weather conditions. There are two main ways to utilize RS for yield assessment. One approach is to directly derive predictions solely from the remote measures. Another approach involves using remotely sensed parameters to enhance computer simulations of crop development and growth. These parameters can be used to validate model output during the growing season, such as biomass or GLAI. They can also be used in a feedback loop to adjust the model's initial conditions or processes (Schepers & Francis, 1998).

2.6.1. Temporal RS Models

Two main types of empirical models have been developed for predicting crop yield: reflectance-based models, which focus on green leaf area or biomass, and thermal-based models, which focus on stress levels. The former was largely based on early research conducted by NASA and university scientists studying grasses, corn, soybeans, wheat, and alfalfa.

Studies conducted by Idso et al. (1977c); Pinter et al. (1981); and Aase and Siddoway (1981) took into account canopy albedo data or VI (Vis) throughout the growing season. They believed that this approach was comparable to the leaf area duration methods commonly employed by crop management experts to forecast final yields. It seems that these empirical approaches may vary depending on the specific variety, as Hatfield (1981) discovered in his study of 82 different varieties of wheat. He found no consistent relationship between yield and the spectral indices. In a study conducted by Aase and Siddoway (1981), it was noted that the correlation between spectral indices and crop yield is influenced by the typical grain-filling conditions for the crop. Any variations in agronomic practices, weather, or soil that deviate from the norm may not always be accurately represented in a straightforward VI-time trajectory. Idso et al. (1980) made an intriguing discovery regarding the relationship between the yields of barley cultivars and spring wheat and the rate of crop senescence, as indicated by the decline in NDVI at the end of the season. The cultivars with higher yields exhibited a more rapid rate of senescence.

Several preliminary investigations linked wheat yields to the temporal trajectories of TIR water stress indices. The crops that experienced higher water stress levels throughout the growing season tended to have the greatest cumulative thermal stress indices as well as produced lower yields. In Hatfield's study (1983b), a more advanced approach was taken by combining regular spectral reflectance and thermal observations to accurately predict yields in grain sorghum (*Sorghum vulgare* L. Moench.) and wheat. This method has proven to be an effective tool for estimating crop yield, with errors of less than 10%. These errors are similar to what is typically observed in recurred small samples taken from large fields (Everitt et al., 1991). This method, although effective, necessitated daily measurements of TIR throughout the grain-filling period to assess crop stress caused by factors such as agronomic practices like nutrients, and soil water and the existing satellite sensor systems do not possess the necessary spatial or temporal resolution to fulfill this data re□uirement. However, this study has demonstrated the effectiveness of combining information from various regions of the spectrum as a potent method for yield prediction (Raun et al., 2001).

2.6.2. Advantages of RS Over Yield Monitors

Combine-mounted yield monitors are now the de facto method for identifying zones for precision crop management and evaluating within-field variability for a wide variety of crops. However, there is an increasing pool of evidence suggesting that yield maps generated from combines may not accurately represent the spatial distribution of plant yields in a field and often do not capture the full extent of variability. Similarly, the ability to identify or manage a particular stress that reduces crop yield is restricted by the end-of-season maps generated by yield monitors (Haas et al., 1975). With the advancements in aircraft-based sensor systems, scientists now have the opportunity to gather data multiple times during the season, thanks to improved spatial and spectral resolution. This has led to the use of remotely sensed imagery as a substitute for a yield map generated by a combine. Having pre-harvest plant estimates productivity allows growers to identify management zones and make more informed marketing decisions in advance. With pre-harvest imagery, growers can easily scout their fields and accurately identify areas of stress. This allows them to promptly diagnose any issues and take remedial actions (Kustas et al., 1990).

Typically, the accuracy of using imagery to estimate yields decreases as the time before harvest increases. This is due to the increased likelihood of factors such as disease, insect infestation, nutrient deficiency, and droughts affecting the yield. For instance, Shanahan et al. (2001) demonstrated that estimating yield based on the duration of corn pollination is not reliable. This is because various crop stresses can lead to variations in tassel emergence dates. In their study, the authors found that the imagery ac□uired during the grain fill period showed a strong relationship ($r_2 > 0.80$) between various VI and grain yield. In a study conducted by Yang et al. (1999), they discovered comparable findings for sorghum ($r_2 = 0.79$). Estimating yields of rain-fed crops using RS can be challenging due to the potentially irreversible loss in yield caused by water stress during critical growth stages. Based on the findings of the Large Area Crop Inventory Experiment, it is evident that collecting imagery multiple times during the season can enhance the accuracy of yield predictions (Wanjura & Hatfield, 1986).

Producers can expect that imagery and yield maps will display similar patterns, but the statistical relationships between yield values extracted from a combine-generated yield map and imagery values are often not very strong. It can be challenging to directly compare the two techniques due to the complexities of grain flow dynamics within the yield monitor and the direction of combined travel. Even when pixel averaging techniques are employed and the combined travel is properly accounted for, r_2 values below 0.25 are common until the field has an extreme range in yield values. It is evident from side-by-side yield comparisons between large yield monitors and small plot combines (Arslan & Colvin, 2002) or from random hand-harvested plots throughout management zones that yield monitor errors can unfairly and poorly reflect on the usefulness of RS as a management tool. An intriguing concept being explored for potential commercial use involves the integration of yield monitor data and pre-harvest imagery.

2.6.3. Integrating RS with Crop Simulation Models

While advancements in simulating crop growth and development have significantly improved, accurately predicting the impact of management factors, unpredictable weather events, and pest pressures on nutrient and crop water needs and final yields remains a challenging task. Remotely sensed imagery has become a practical tool for providing crop simulation models with dynamic canopy state variables in both space and time (Wiegand et al., 1979). With the help of crop models, the information obtained from remotely sensed images can be enhanced. These models can extrapolate data for times when inclement weather hinders data collection and predict how crops and yields will respond to different management strategies.

Moulin et al. (1998) conducted a review on integrating remotely sensed data into crop models, exploring various approaches. While the goal of these integrated approaches has often been to track crop yield and condition at state, county, and regional scales, more recent work has focused on forecasting within-field variability in crop status (Sadler et al., 2002). The models can be directly coupled with remotely sensed imagery through the use of biomass, green leaf area index (GLAI), and phenological stages; alternatively, they can be indirectly coupled by inferring factors such as weed and insect pressure, diseases, nutrients, water status, and fAPAR. Examples include:

i. Iteratively adjusting the model's cultivar-specific parameters and the model's initial conditions to ensure that the model's predictions align with periodic remotely sensed estimates of ET and LAI (Maas, 1993).
ii. Forcing model predictions to align with remotely sensed estimates of real field conditions during a specific point in the season.
iii. Using radiative transfer models to enable direct comparison between satellite reflectance data and a crop model's predictions.

Crop models enable the simulation of various management strategies in varying weather scenarios. Additionally, the models can account for spatial variability and offer sporadic "reality checks" using remotely sensed data. As these methods develop further, it will be more and more crucial to integrate model output with decision support systems that take into account various factors like labor, economic, and time constraints (Jones, 2000).

Effective management of the huge quantity of remotely sensed along with additional data within a GIS will require the use of decision support systems (Clothier et al., 1986).

2.7. OTHER ASPECTS OF CROP MANAGEMENT

2.7.1. Plant Population

Plant density is a crucial variable for optimizing productivity in various systems. Populations can fluctuate due to factors such as soil parameters, planter performance, weather conditions, field slope, and aspect, as well as seedling disease. When population numbers for some crops, such as corn or non-tillering sorghum grain varieties, are outside of relatively narrow optimum ranges, yield potential is diminished (Pinter et al., 1979).

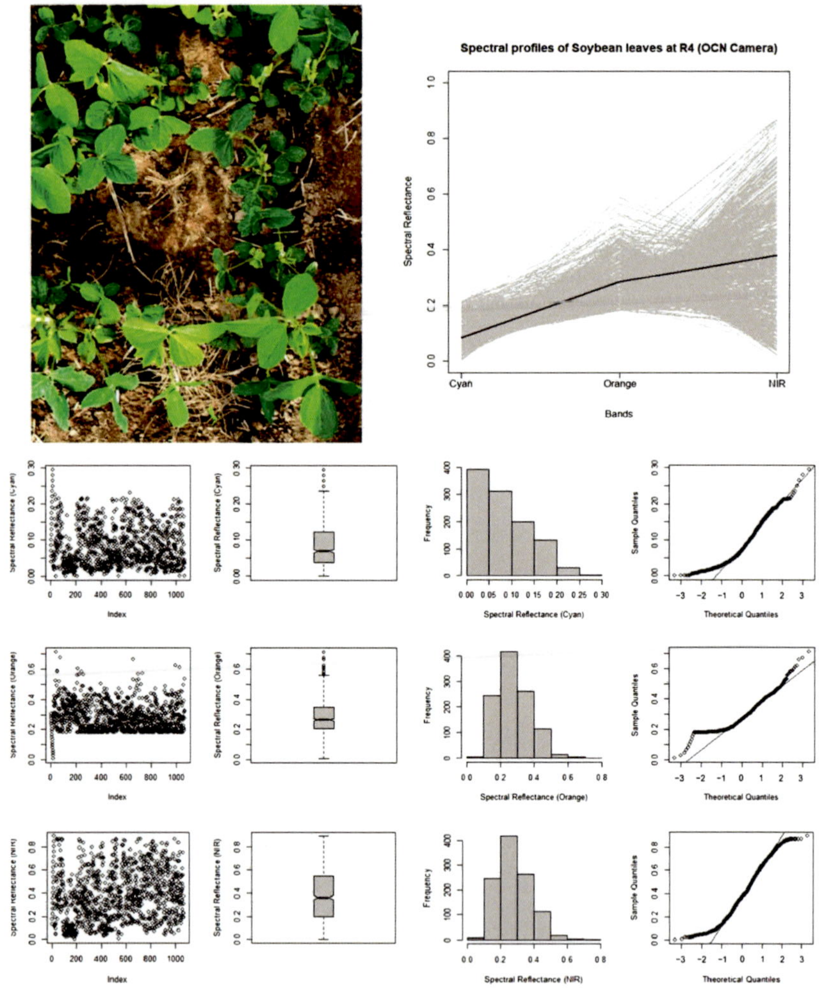

Figure 2.6. *The average OCN spectral profile and exploratory data analysis for the R4 stage are presented. The upper left panel shows an example of the R4 stage. The upper right panel displays the spectral profiles of 1,000 points from the OCN cam-*

era (grey lines) and the average spectral profile of the reproductive stage (black line). The lower part of the image includes the results of exploratory data analysis for each band (scatter plot (index plot), box plot, histogram, and theoretical quantiles, respectively).

Source: Vásquez, R. A. R., Heenkenda, M. K., Nelson, R., & Segura Serrano, L. (2023). Developing a new vegetation index using cyan, orange, and near-infrared bands to analyze soybean growth dynamics. Remote Sensing, 15(11), 2888. https://doi.org/10.3390/rs15112888.

Depending on the crop (wheat, cotton, soybeans, etc.), tillering or branching makes final yields less dependent on population density. However, early canopy closure and uniform stand emergence are good ways to control weeds and affect early maturity.

On the other hand, if the stand is too dense, it can lead to plants that do not produce marketable fruit or have a canopy that is more attractive to pests or prone to diseases. Now, with variable-rate planters, seeding rates can be adjusted to account for variations in emergence or to reach densities that are more closely matched to the soil properties unique to a field's site (Condit, 1970) (Figure 2.6).

Commercial RS providers aim to provide accurate and timely plant density estimates for growers. These estimates can help identify insect infestations or seeding diseases, determine the need for replanting, plan herbicide and fertilizer requirements, and interpret end-of-season yield maps. Plattner & Hummel (1996) developed a non-contact, combined-mounted sensor that collected data on corn plant population, skips, doubles, and spacing using a photoelectric emitter and detector pair.

At the early growth stage, the sensor provided an estimated plant spacing with a 3% error, which increased to 6% at harvest. During field tests, the filtering algorithms successfully eliminated the disruptions caused by small weeds, but large corn leaves remained a source of errors (Upchurch & Mahan, 1988).

For management purposes, plant populations could ideally be determined using multispectral imagery captured soon after emergence. In reality, the plants are typically small in size, and their signal gets overshadowed by the soil.

Obtaining images in the early morning or from different angles can help with identifying plants with low leaf area levels (Pearson et al., 1976). As more sensitive sensors are utilized and improved techniques are utilized to calibrate them and get rid of the effects of changes in the soil background (Moran et al., 1994), it should become easier to get a good idea of how dense plants are in the early season.

2.7.2. Growth Regulators and Defoliants

Growers are increasingly utilizing chemical plant growth regulators, such as mepiquat chloride (Pix), to manipulate plant growth, making mechanical harvesting easier and promoting early maturity. During the late 1970s, ARS scientists utilized aerial CIR photography to assess the impact of defoliants on late-season fruiting and the population of overwintering pink bollworms (*Pectinophora gossypiella* Saunders; Henneberry et al., 1979). Later on, Richardson & Gausman (1982) conducted a study on the impact of Pix on the reflectance properties of cotton leaves and canopies. Their findings showcased the promising use of RS to assess large areas of treated cotton.

In their study, Shanahan & Nielsen (1987) utilized the CWSI to assess the effectiveness of plant growth regulators in conserving water usage during the early stages of corn growth in semiarid regions. Researchers in Mississippi utilized high spatial resolution, multispectral imagery to effectively apply Pix in just the required areas to control rank growth in over 400 hectares of cotton (Qi, 1994).

2.8. OVERALL CHALLENGES AND OPPORTUNITIES

In a groundbreaking essay from over 20 years ago, Ray Jackson emphasized the crucial factors for farmers when it comes to utilizing RS technology for farm management decisions. According to Jackson, the most significant aspects are the timeliness, frequency, and spatial resolution of the data, in that specific order. His observations continue to hold in the present. There have been significant advancements in the field of view, but unfortunately, most farmers still lack regular access to images of their farms. Even if they do have access, the slow processing time remains a persistent issue (Oliva et al., 1994).

Although there are some limitations, it is undeniable that RS technologies are going to play a significant role in various aspects of farming in the future. As more affordable products with higher temporal and spatial resolution become available, grower acceptance is expected to increase. By implementing these measures, costs can be reduced, resulting in improved coverage and faster image delivery. To instill confidence in growers, RS providers must convert imagery to reflectance, prioritize calibration issues, and establish standardized wavelengths and data collection techniques. This will produce a more dependable product that delivers the same story every year and provides valuable information to farming operations as a whole (Nielsen, 1990).

From a research perspective, several interconnected and overarching challenges need to be addressed to advance RS beyond its current qualitative applications for better crop management. One aspect involves gaining a deep understanding and the ability to accurately model bidirectional reflectance properties of the agricultural targets. The simplest relationship between NDVI and green leaf area index changes a lot depending on the solar illumination angles, and the orientation of the plant rows or the sensor viewing direction. Understanding the bidirectional effects in crop management allows for a more accurate analysis of spectral characteristics, reducing their reliance on factors such as time of day, season, row direction, or spacing (McMurtrey et al., 1996).

Another significant research challenge involves the development of stress detection algorithms that can consistently perform reliably over different spaces and times. Techniques need to be applicable regardless of the location, soil types, and management factors. They should also perform optimally throughout the entire growing season, from the initial planting to the final stage of maturity (McGwire et al., 2000). Here, it is important to find distinct indicators for particular stresses in the ever-changing context of typical crop development and growth, such as the spectral complexities brought on by insufficient plant cover. With the help of advanced techniques like spectral mixing analysis, it is possible to differentiate stress signals caused by water, nutrients, and pests from other factors like soil and non-plant elements (Kimball et al., 1999). Advanced approaches will combine expert and decision support systems with remotely sensed parameters to compare temporal and spatial crop spectra patterns and emittance with historical data, all while taking current weather and management practices into account. By integrating remote observations with current crop simulation models, the models will gain a spatial component that will enhance their predictive power and increase their utility for farm managers (McFarlane et al., 1980). The vast amount of spatial, temporal, and spectral data that can be extracted from a series of remotely sensed images presents exceptional possibilities for managing and managing agricultural resources on a local and global level. With a deep understanding of crop management, one can combine historic imagery with crop calendars, precipitation records, heat units, and yield monitor data to create maps that highlight areas susceptible to water stress, pest issues, or nutrient deficiency in specific environmental conditions (Kustas et al., 1990). Decision support systems could then use current imagery to give an early warning of yield, thereby lowering stress. As variable rate technology becomes more common, producers would find this information extremely helpful when making decisions. Archived satellite imagery allows policymakers and scientists to closely monitor the effects of global change on agriculture worldwide. For growers looking to assess their environmental stewardship or earn credits for carbon sequestration, imagery can be a valuable tool to document their achievements in a fair and scientifically based manner (Mahan & Upchurch, 1988).

2.9. SUMMARY

Managing agricultural resources in today's world is a highly intricate task that is being enhanced by the integration of cutting-edge advancements in geographic positioning technologies, information sciences, and RS systems. Extensive research on the spectral properties of soils and crops, and their relationship to biophysical and agronomic parameters, has been conducted by ARS researchers in collaboration with NASA and university scientists. This collaborative effort has spanned several programs over the last 40 years. Multiple areas of crop management are already seeing improvements thanks to the use of RS technology. As growers become more familiar with its use, they will discover new opportunities. The future holds immense potential for the integration of the spatially and temporally rich information obtained from remotely sensed multi- and hyper-spectral imagery with the capabilities of management-oriented crop simulation models.

REFERENCES

1. Aase, J. K., & Siddoway, F. H. (1981). Assessing winter wheat dry matter production via spectral reflectance measurements. *Remote Sensing of Environment, 11,* 267–277.
2. Adams, M. L., Norvell, W. A., Peverly, J. H., & Philpot, W. D. (1993). Fluorescence and reflectance characteristics of manganese deficient soybean leaves: Effects of leaf age and choice of leaflet. In *Plant Nutrition—from Genetic Engineering to Field Practice: Proceedings of the Twelfth International Plant Nutrition Colloquium, 21–26 September 1993, Perth, Western Australia,* 156 (pp. 261–264).
3. Adams, M. L., Norvell, W. A., Philpot, W. D., & Peverly, J. H. (2000). Spectral detection of micronutrient deficiency in 'Bragg' soybean. *Agronomy Journal, 92*(2), 261–268.
4. Adamsen, F. J., Pinter Jr, P. J., Barnes, E. M., LaMorte, R. L., Wall, G. W., Leavitt, S. W., & Kimball, B. A. (1999). Measuring wheat senescence with a digital camera. *Crop Science, 39*(3), 719–724.
5. Alderfasi, A. A., & Nielsen, D. C. (2001). Use of crop water stress index for monitoring water status and scheduling irrigation in wheat. *Agricultural Water Management, 47*(1), 69–75.
6. Allen, R. G., Pereira, L. S., Raes, D., & Smith, M. (1998). *Crop evapotranspiration—Guidelines for computing crop water requirements—FAO Irrigation and drainage paper 56.* FAO, Rome, 300(9), D05109.
7. Arslan, S., & Colvin, T. S. (2002). Grain yield mapping: Yield sensing, yield reconstruction, and errors. *Precision Agriculture, 3,* 135–154.
8. Barnes, E. M., Sudduth, K. A., Hummel, J. W., Lesch, S. M., Corwin, D. L., Yang, C., & Bausch, W. C. (2003). Remote- and ground-based sensor techniques to map soil properties. *Photogrammetric Engineering & Remote Sensing, 69*(6), 619–630.
9. Barnes, W. L., Dereux, A., & Ebbesen, T. W. (2003). Surface reeban subwavelength optics. *Nature, 424*(6950), 824–830.
10. Bartholic, J. F., Namken, L. N., & Wiegand, C. L. (1972). Aerial thermal scanner to determine temperatures of soils and crop canopies differing in water stress. *Agronomy Journal, 64*(5), 603–608.
11. Bauer, M. E. (1975). The role of remote sensing in determining the distribution and yield of crops. *Advances in Agronomy, 27,* 271–304.
12. Bausch, W. C., & Diker, K. (2001). Innovative remote sensing techniques to increase nitrogen use efficiency of corn. *Communications in Soil Science and Plant Analysis, 32*(7–8), 1371–1390.

13. Beerwinkle, K. R., Witz, J. A., & Schleider, P. G. (1993). An automated, vertical-looking, X-band radar system for continuously monitoring aerial insect activity. *Transactions of the ASAE, 36*(3), 965–970.
14. Ben-Asher, J., Phene, C. J., & Kinarti, A. (1992). Canopy temperature to assess daily evapotranspiration and management of high-frequency drip irrigation systems. *Agricultural Water Management, 22*(4), 379–390.
15. Blackmer, T. M., Schepers, J. S., Varvel, G. E., & Walter-Shea, E. A. (1996). Nitrogen deficiency detection using reflected shortwave radiation from irrigated corn canopies. *Agronomy Journal, 88*(1), 1–5.
16. Bowman, W. D. (1989). The relationship between leaf water status, gas exchange, and spectral reflectance in cotton leaves. *Remote Sensing of Environment, 30*(3), 249–255.
17. Burke, J. J. (1994, August). Approaches to integrated water resource development and management: The Kafue Basin, Zambia. In *Natural Resources Forum,* Vol. 18, No. 3, (pp. 181–192).
18. Carlson, T. N., Taconet, O., Vidal, A., Gillies, R. R., Olioso, A., & Humes, K. (1995). An overview of the workshop on thermal remote sensing held at La Londe les Maures, France, September 20–24, 1993. *Remote Sensing Reviews, 12*(3–4), 147–158.
19. Carter, G. A. (1991). Primary and secondary effects of water content on the spectral reflectance of leaves. *American Journal of Botany, 78*(7), 916–924.
20. Chappelle, E. W., Wood, F. M., McMurtrey, J. E., & Newcomb, W. W. (1984). Laser-induced fluorescence of green plants. 1: A techni□ue for the remote detection of plant stress and species differentiation. *Applied Optics, 23*(1), 134–138.
21. Choudhury, B. J., Ahmed, N. U., Idso, S. B., Reginato, R. J., & Daughtry, C. S. (1994). Relations between evaporation coefficients and vegetation indices studied by model simulations. *Remote Sensing of Environment, 50*(1), 1–17.
22. Clarke, T. R. (1997). An empirical approach for detecting crop water stress using multispectral airborne sensors. *Precision Agriculture, 7*(1), 9–16.
23. Clawson, K. L., Jackson, R. D., & Pinter Jr, P. J. (1989). Evaluating plant water stress with canopy temperature differences. *Agronomy Journal, 81*(6), 858–863.
24. Clothier, B. E., Clawson, K. L., Pinter Jr, P. J., Moran, M. S., Reginato, R. J., & Jackson, R. D. (1986). Estimation of soil heat flux from net radiation during the growth of alfalfa. *Agricultural and Forest Meteorology, 37*(4), 319–329.

25. Condit, H. R. (1970). The spectral reflectance of American soils. *Photogrammetric Engineering, 36*(9), 955–966.
26. Cook, C. G., Escobar, D. E., Everitt, J. H., Cavazos, I., Robinson, A. F., & Davis, M. R. (1999). Utilizing airborne video imagery in Kenaf management and production. *Industrial Crops and Products, 9*(3), 205–210.
27. Corp, L. A., McMurtrey, J. E., Chappelle, E. W., Daughtry, C. S., & Kim, M. S. (1997). UV band fluorescence (in vivo) and its implications for the remote assessment of nitrogen supply in vegetation. *Remote Sensing of Environment, 61*(1), 110–117.
28. Daughtry, C. S. (2001). Discriminating crop residues from the soil by shortwave infrared reflectance. *Agronomy Journal, 93*(1), 125–131.
29. Dickson, M. A., Bausch, W. C., & Howarth, M. S. (1994). Classification of a broadleaf weed, a grassy weed, and corn using image processing techniques. *Transactions of the ASAE, 2345,* 297–305.
30. Donald, M. (2000). The central role of culture in cognitive evolution: A reflection on the myth of the isolated mind. *Culture, Thought, and Development, 2*(1), 19–38.
31. Dupont, J. K., Willers, J. L., Campanella, R., Seal, M. R., & Hood, K. B. (2000, November). Spatially variable insecticide applications through remote sensing. In *Proceedings of the Beltwide Cotton Conference,* 1, 426–429.
32. Ehrler, W. L. (1973). Cotton leaf temperatures as related to soil water depletion and meteorological factors. *Agronomy Journal, 65*(3), 404–409.
33. Elliott, N. C., Kieckhefer, R. W., Lee, J. H., & French, B. W. (1999). Influence of within-field and landscape factors on aphid predator populations in wheat. *Landscape Ecology, 14*(3), 239–252.
34. Everitt, J. H., Escobar, D. E., Villarreal, R., Noriega, J. R., & Davis, M. R. (1991). Airborne video systems for agricultural assessment. *Remote Sensing of Environment, 35*(2–3), 231–242.
35. Feldhake, C. M., & Edwards, W. M. (1992). Pasture canopy temperature under cloudy humid conditions. *Agricultural and Forest Meteorology, 60*(1–2), 133–144.
36. Fiscus, E. L., Mahbub-Ul Alam, A. N. M., & Hirasawa, T. (1991). Fractional integrated stomatal opening to control water stress in the field. *Crop Science, 31*(4), 1001–1008.
37. Franz, E., Gebhardt, M. R., & Unklesbay, K. B. (1991). The use of local spectral properties of leaves as an aid for identifying weed seedlings in digital images. *Transactions of the ASAE, 34*(2), 682–687.

38. Gamon, J., Serrano, L., & Surfus, J. S. (1997). The photochemical reflectance index: An optical indicator of photosynthetic radiation use efficiency across species, functional types, and nutrient levels. *Oecologia, 112,* 492–501.
39. Gardner, B. R., Nielsen, D. C., & Shock, C. C. (1992). Infrared thermometry and the crop water stress index. II. Sampling procedures and interpretation. *Journal of Production Agriculture, 5*(4), 466–475.
40. Gausman, H. W. (1974). Leaf reflectance of near-infrared. *Photogrammetric Engineering, 40*(2), 183–191.
41. Gausman, H. W., Menges, R. M., Richardson, A. J., Walter, H., Rodriguez, R. R., & Tamez, S. (1981). Optical parameters of leaves of seven weed species. *Weed Science, 29*(1), 24–26.
42. Haas, R. H., Deering, D. W., Rouse, J. W., & Schell, J. A. (1975, June). Monitoring vegetation conditions from Landsat for use in range management. In *NASA Earth Resources Survey Symposium Proc., Houston, Texas,* Vol. 1 (pp. 43–52).
43. Hanks, J. E., & Beck, J. L. (1998). Sensor-controlled hooded sprayer for row crops. *Weed Technology, 12*(2), 308–314.
44. Hart, M. D., Meyers Jr, J. J., Wood, Z. A., Nakakita, T., Applegate, J. C., Erickson, N. R., & Barybin, M. V. (2021). Tuning π-acceptor/σ-donor ratio of the 2-isocyanoazulene ligand: Non-fluorinated rival of pentafluorophenyl isocyanide and trifluorovinyl isocyanide discovered. *Molecules, 26*(4), 981.
45. Hart, W. G., & Myers, V. I. (1968). Infrared aerial color photography for detection of populations of brown soft scale in citrus groves. *Journal of Economic Entomology, 61*(3), 617–624.
46. Hatfield, J. L. (1983). Remote sensing estimators of potential and actual crop yield. *Remote Sensing of Environment, 13*(4), 301–311.
47. Hatfield, J. L., Burke, J. J., Mahan, J. R., & Wanjura, D. F. (1987). Foliage temperature measurements: A link between the biological and physical environment. *Agronomy Journal, 79,* 99–102.
48. Henneberry, T. J., Bariola, L. A., Hart, W., Davis, M., Ingle, S., Kittock, D., & Arle, H. (1979). Parameters of cotton cultivation from infrared aerial photography. *Photogrammetric Engineering and Remote Sensing, 45*(8), 1129–1133.
49. Hickman, M. V., Everitt, J. H., Escobar, D. E., & Richardson, A. J. (1991). Aerial photography and videography for detecting and mapping dicamba injury patterns. *Weed Technology, 5*(4), 700–706.

50. Hunt Jr, E. R., Everitt, J. H., Ritchie, J. C., Moran, M. S., Booth, D. T., Anderson, G. L., & Seyfried, M. S. (2003). Applications and research using remote sensing for rangeland management. *Photogrammetric Engineering & Remote Sensing, 69*(6), 675–693.
51. Hutmacher, R. B., Steiner, J. J., Vail, S. S., & Ayars, J. E. (1991). Crop water stress index for seed alfalfa: Influences of within-season changes in plant morphology. *Agricultural Water Management, 19*(2), 135–149.
52. Idso, S. B. (1982). Non-water-stressed baselines: A key to measuring and interpreting plant water stress. *Agricultural Meteorology, 27*(1–2), 59–70.
53. Jackson, R. D. (1982). Canopy temperature and crop water stress. In *Advances in Irrigation,* Vol. 1, (pp. 43–85).
54. Jones, D., & Barnes, E. M. (2000). Fuzzy composite programming to combine remote sensing and crop models for decision support in precision crop management. *Agricultural Systems, 65*(3), 137–158.
55. Kimball, B. A., LaMorte, R. L., Pinter Jr, P. J., Wall, G. W., Hunsaker, D. J., Adamsen, F. J., & Brooks, T. J. (1999). Free-air CO enrichment and soil nitrogen effects on energy balance and evapotranspiration of wheat. *Water Resources Research, 35*(4), 1179–1190.
56. Kustas, W. P., Moran, M. S., Jackson, R. D., Gay, L. W., Duell, L. F. W., Kunkel, K. E., & Matthias, A. D. (1990). Instantaneous and daily values of the surface energy balance over agricultural fields using remote sensing and a reference field in an arid environment. *Remote Sensing of Environment, 32*(2–3), 125–141.
57. Maas, S. J. (1988). Use of remotely sensed information in agricultural crop growth models. *Ecological Modelling, 41*(3–4), 247–268.
58. Mahan, J. R., & Upchurch, D. R. (1988). Maintenance of constant leaf temperature by plants—I. Hypothesis-limited homeothermy. *Environmental and Experimental Botany, 28*(4), 351–357.
59. Masoni, A., Ercoli, L., & Mariotti, M. (1996). Spectral properties of leaves deficient in iron, sulfur, magnesium, and manganese. *Agronomy Journal, 88*(6), 937–943.
60. McFarlane, J. C., Watson, R. D., Theisen, A. F., Jackson, R. D., Ehrler, W. L., Pinter, P. J., & Reginato, R. J. (1980). Plant stress detection by remote measurement of fluorescence. *Applied Optics, 19*(19), 3287–3289.
61. McGwire, K., Minor, T., & Fenstermaker, L. (2000). Hyperspectral mixture modeling for quantifying sparse vegetation cover in arid environments. *Remote Sensing of Environment, 72*(3), 360–374.

62. McMurtrey III, J. E., Chappelle, E. W., Kim, M. S., Corp, L. A., & Daughtry, C. S. (1996). Blue-green fluorescence and visible-infrared reflectance of corn (*Zea mays* L.) grain for in situ field detection of nitrogen supply. *Journal of Plant Physiology, 148*(5), 509–514.

63. McMurtrey III, J. E., Chappelle, E. W., Kim, M. S., Meisinger, J. J., & Corp, L. A. (1994). Distinguishing nitrogen fertilization levels in field corn (*Zea mays* L.) with actively induced fluorescence and passive reflectance measurements. *Remote Sensing of Environment, 47*(1), 36–44.

64. Millard, J. P., Reginato, R. J., Idso, S. B., Jackson, R. D., Goettelman, R. C., & Leroy, M. J. (1980). Experimental relations between airborne and ground-measured wheat canopy temperatures. *Photogrammetric Engineering and Remote Sensing, 46*(2), 221–224.

65. Moran, M. S., Clarke, T. R., Inoue, Y., & Vidal, A. (1994). Estimating crop water deficit using the relation between surface-air temperature and spectral vegetation index. *Remote Sensing of Environment, 49*(3), 246–263.

66. Moran, S., Fitzgerald, G., Rango, A., Walthall, C., Barnes, E., Bausch, W., & Upchurch, D. (2003). Sensor development and radiometric correction for agricultural applications. *Photogrammetric Engineering & Remote Sensing, 69*(6), 705–718.

67. Moulin, S., Bondeau, A., & Delecolle, R. (1998). Combining agricultural crop models and satellite observations: From field to regional scales. *International Journal of Remote Sensing, 19*(6), 1021–1036.

68. Nagler, P. L., Daughtry, C. S. T., & Goward, S. N. (2000). Plant litter and soil reflectance. *Remote Sensing of Environment, 71*(2), 207–215.

69. Nielsen, D. C. (1990). Scheduling irrigations for soybeans with the crop water stress index (CWSI). *Field Crops Research, 23*(2), 103–116.

70. Nouvellon, Y., Moran, M. S., Seen, D. L., Bryant, R., Rambal, S., Ni, W., & Qi, J. (2001). Coupling a grassland ecosystem model with Landsat imagery for a 10-year simulation of carbon and water budgets. *Remote Sensing of Environment, 78*(1–2), 131–149.

71. Oliva, R. N., Steiner, J. J., & Young III, W. C. (1994). White clover seed production: II. Soil and plant water status on yield and yield components. *Crop Science, 34*(3), 768–774.

72. Osborne, S. L., Schepers, J. S., Francis, D. D., & Schlemmer, M. R. (2002). Use of spectral radiance to estimate in-season biomass and grain yield in nitrogen- and water-stressed corn. *Crop Science, 42*(1), 165–171.

73. Pearson, R. L., Miller, L. D., & Tucker, C. J. (1976). Hand-held spectral radiometer to estimate gramineous biomass. *Applied Optics, 15*(2), 416–418.

74. Peñuelas, J., Gamon, J. A., Fredeen, A. L., Merino, J., & Field, C. B. (1994). Reflectance indices associated with physiological changes in nitrogen- and water-limited sunflower leaves. *Remote Sensing of Environment, 48*(2), 135–146.

75. Peynado, A., Gausman, H. W., Escobar, D. E., & Rodriguez, R. R. (1980). Detecting freeze damage to citrus leaves. *HortScience, 15,* 435.

76. Pinter Jr, P. J. (1986). Effect of dew on canopy reflectance and temperature. *Remote Sensing of Environment, 19*(2), 187–205.

77. Pinter Jr, P. J., Hatfield, J. L., Schepers, J. S., Barnes, E. M., Moran, M. S., Daughtry, C. S., & Upchurch, D. R. (2003). Remote sensing for crop management. *Photogrammetric Engineering & Remote Sensing, 69*(6), 647–664.

78. Pinter Jr, P. J., Stanghellini, M. E., Reginato, R. J., Idso, S. B., Jenkins, A. D., & Jackson, R. D. (1979). Remote detection of biological stresses in plants with infrared thermometry. *Science, 205*(4406), 585–587.

79. Plattner, C. E., & Hummel, J. W. (1996, January). Corn plant population sensor for precision agriculture. In *Proceedings of the Third International Conference on Precision Agriculture,* Vol. 2 (pp. 785–794).

80. Qi, J., Chehbouni, A., Huete, A. R., Kerr, Y. H., & Sorooshian, S. (1994). A modified soil-adjusted vegetation index. *Remote Sensing of Environment, 48,* 119–126.

81. Raun, W. R., Solie, J. B., Johnson, G. V., Stone, M. L., Lukina, E. V., Thomason, W. E., & Schepers, J. S. (2001). In-season prediction of potential grain yield in winter wheat using canopy reflectance. *Agronomy Journal, 93*(1), 131–138.

82. Rhoades, J. D., Manteghi, N. A., Shouse, P. J., & Alves, W. J. (1989). Soil electrical conductivity and soil salinity: New formulations and calibrations. *Soil Science Society of America Journal, 53*(2), 433–439.

83. Richardson, A. J., & Gausman, H. W. (1982). Reflectance differences between untreated and mepi□uat chloride-treated, field-grown cotton through a growing season. *Remote Sensing of Environment, 12*(6), 501–507.

84. Sadler, K., & Tam, J. P. (2002). Peptide dendrimers: Applications and synthesis. *Reviews in Molecular Biotechnology, 90*(3–4), 195–229.

85. Schepers, J. S., & Francis, D. D. (1998). Precision agriculture—What's in our future? *Communications in Soil Science and Plant Analysis, 29*(11–14), 1463–1469.

86. Shanahan, J. F., & Nielsen, D. C. (1987). Influence of growth retardants (anti-gibberellins) on corn vegetative growth, water use, and grain yield under different levels of water stress. *Agronomy Journal, 79*(1), 103–109.

87. Shanahan, J. F., Schepers, J. S., Francis, D. D., Varvel, G. E., Wilhelm, W. W., Tringe, J. M., & Major, D. J. (2001). Use of remote-sensing imagery to estimate corn grain yield. *Agronomy Journal, 93*(3), 583–589.

88. Sudduth, K. A., & Hummel, J. W. (1993). Soil organic matter, CEC, and moisture sensing with a portable NIR spectrophotometer. *Transactions of the ASAE, 36*(6), 1571–1582.

89. Tian, Q., & Reed, J. W. (1999). Control of auxin-regulated root development by the *Arabidopsis thaliana* SHY2/IAA3 gene. *Development, 126*(4), 711–721.

90. Upchurch, D. R., & Mahan, J. R. (1988). Maintenance of constant leaf temperature by plants—II. Experimental observations in cotton. *Environmental and Experimental Botany, 28*(4), 359–366.

91. Wanjura, D. F., & Hatfield, J. L. (1986). PAR and IR reflectance, transmittance, and absorptance of four crop canopies. *Transactions of the ASAE, 29*(1), 143–150.

92. Wiegand, C. L., Richardson, A. J., & Kanemasu, E. T. (1979). Leaf area index estimates for wheat from LANDSAT and their implications for evapotranspiration and crop modeling. *Agronomy Journal, 71*(2), 336–342.

93. Willer, D. O., McFadden, G., & Evans, D. H. (1999). The complete genome se□uence of Shope (rabbit) fibroma virus. *Virology, 264*(2), 319–343.

94. Yang, C., & Anderson, G. L. (1999). Airborne videography to identify spatial plant growth variability for grain sorghum. *Precision Agriculture, 1*, 67–79.

Chapter **3**

REMOTE SENSING IN CROP DISEASE DETECTION AND MANAGEMENT

Contents

3.1. INTRODUCTION

Remote sensing (RS) technologies have the potential to greatly assist in a wide range of environmental management and conservation tasks. However, it is crucial to recognize the limitations just as much as the benefits. When remotely sensed data is used correctly, it can greatly enhance knowledge and provide valuable insights. However, when used improperly, it can lead to costly mistakes and produce misleading results (Yang, 2020).

Remote sensing plays a crucial role in managing crops and detecting diseases in modern agriculture. With the help of advanced technology such as ground sensors, drones, and satellites, remote sensing empowers farmers to effortlessly monitor the health of their crops across vast expanses, eliminating the need for manual field inspections (Qin & Zhang, 2005). This technology is incredibly valuable in identifying diseases at an early stage, enabling farmers to take prompt action and effectively contain their spread. Remote sensing can identify issues that may not be apparent to the human eye by analysis of data using different wavelengths of light. This aids farmers in applying treatments more efficiently, conserving resources and time while preserving the environment. Remote sensing plays a crucial role in enhancing crop yields, maintaining a consistent food supply, and promoting sustainable farming practices (Pinter et al., 2003).

3.2. PRACTICAL AND PHYSICAL RESTRICTIONS OF REMOTELY SENSED IMAGERY

It is undeniable that advancements in aerial as well as satellite remote sensing have significantly enhanced the capacity to accurately map land cover and effectively monitor various environmental phenomena. However, it is important to acknowledge the inherent limitations that are sometimes downplayed or disregarded in the professional literature regarding remote sensing. In addition to pointing out some trade-offs between drawbacks and advantages—among which some might not be immediately intuitive—it is also necessary to make you aware of some of the limits. To provide an understanding of how these problems change over time, this book will highlight research and development trends in RS throughout the content (Xie et al., 2008).

Remote sensing has its fair share of limitations. One of them is the difficulty in accurately aligning various imagery with a map base. Additionally, taking beneficial images in a hazy or cloudy condition can be □uite challenging. Another limitation is the ability to record data below water surfaces or vegetative canopies. In addition, there are concerns regarding the practicality of RS imagery and the products derived from it (Salin, 1992). Many such problems can be technical, like the various factors that influence the resolution of the imagery features or those that influence image classification. There are also practical limitations to consider, such as the project's resources like human resources, time, and financial resources alongside location considerations (Watts et al., 2012).

3.2.1. Resolution

Remote sensing discussions often focus on enhancing image resolution and capturing finer details. Various factors influence the levels of detail that can be observed or resolved in imagery. The four categories of image resolution that are typically mentioned are:

- Spatial;
- Spectral;
- Radiometric; and
- Temporal.

The spatial resolution, also known as "resolution," refers to the pixel size in the ground dimensions. The pixel represents the tiniest individual element and unit of display in an image. There is a conventional misunderstanding surrounding spatial resolution, which assumes that a finer resolution is inevitably superior. While several image providers offer high-resolution imagery of less than 1 m, it is important to consider that they may not always be the most suitable option for certain projects. The IKONOS image (1 m) is visually appealing, showcasing distinct houses, and tree crowns among other intricate details which cannot be discerned in the other images (Kühlbrandt, 2014). Then, why would anyone even think about utilizing other pictures?

A high-resolution image needs greater quantities of data or the number of pixels to encompass a specific region, resulting in a smaller overall coverage area compared to a coarse-resolution image. It is crucial to take this into account when performing investigations that encompass a vast region. Creating a single seamless image by combining multiple images that encompass an area of interest can be quite challenging. Typically, it may be advisable to utilize a minimum number of images necessary to represent a study region. In connection with this, data file size is influenced by the spatial resolution. The higher the resolution, the larger the file size becomes. A photograph with a 1 m resolution, for instance, would have 100 times more pixels than the same picture with a 10 m resolution, resulting in a 100 times larger size (Sheldrick & Schneider, 1997).

One drawback of utilizing high-resolution imagery is its potential cost, which can amount to tens of USD per kilometer squared. With the increasing availability of high-resolution imagery and the growing number of providers in the market, it is possible that prices may decrease. However, it is expected that prices will remain relatively high in the near future, which may cause potential buyers to pause before making a purchase (Fogassy et al., 2006).

Another limitation that may not be immediately apparent is the challenge of processing high-resolution images through automated methods. Conventional categorization techniques that operate on a pixel-by-pixel level often produce unsatisfactory results when trying to distinguish distinct elements of a feature of interest (Carleer et al., 2005). For instance, in the IKONOS image shown in Figure 3.1, there is a need to identify the forest class. However, when applying a conventional pixel-by-pixel categorization, the resulting classes are the different elements of a forest, including soil, leaves, stems, and shadow (Figure 3.3). When observing a forest through a coarser spatial resolution, just one pixel will contain a combination of various elements such as the soil, leaves, stems, and shadows. This combination is what is identified as a forest, known as a "mixed pixel." There is a technique called "spectral unmixing" that can be utilized to calculate the proportional abundance of particular elements within a single pixel (Liu & Wu, 2005).

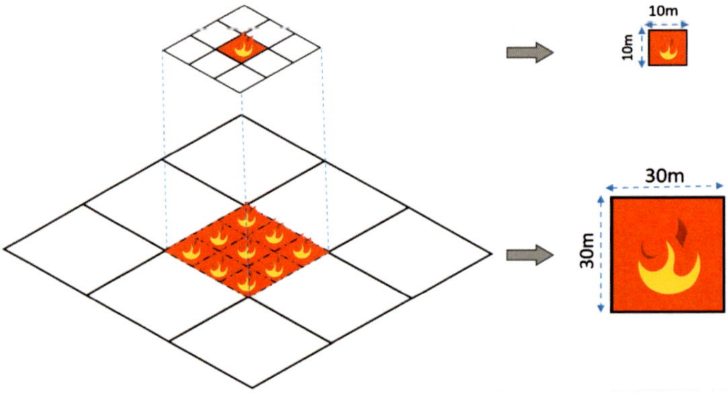

Figure 3.1. *The spatial detail levels vary with the different spatial resolutions.*

Source: Tavakkoli, P. S., Einali, G., Ghorbanzadeh, O., Nachappa, T. G., Gholamnia, K., Blaschke, T., & Ghamisi, P. (2022). A Google Earth Engine approach for wildfire susceptibility prediction fusion with remote sensing data of different spatial resolutions. Remote Sensing, 14(3), 672. https://doi.org/10.3390/rs14030672 (Creative Commons License).

Feature recognition algorithms have been used in automated analysis methods (Meinel & Neubert, 2004) to classify pixels into meaningful groups. These methods can effectively group adjacent features, like the various elements of a tree (branches, leaves, and shadows), into distinct objects such as trees or forests. Nevertheless, the software programs that support these methods can be □uite expensive and re□uire a certain level of technical proficiency to navigate. The perfect imagery in the task of classification may typically have a spatial resolution that is in line with the end map representation based on the data's intended size. Overly fine-resolution imagery will complicate automated classification, whereas overly coarse-resolution imagery will generalize and produce a map unsatisfactory for the user (Nilsson, 1995) (Figure 3.2).

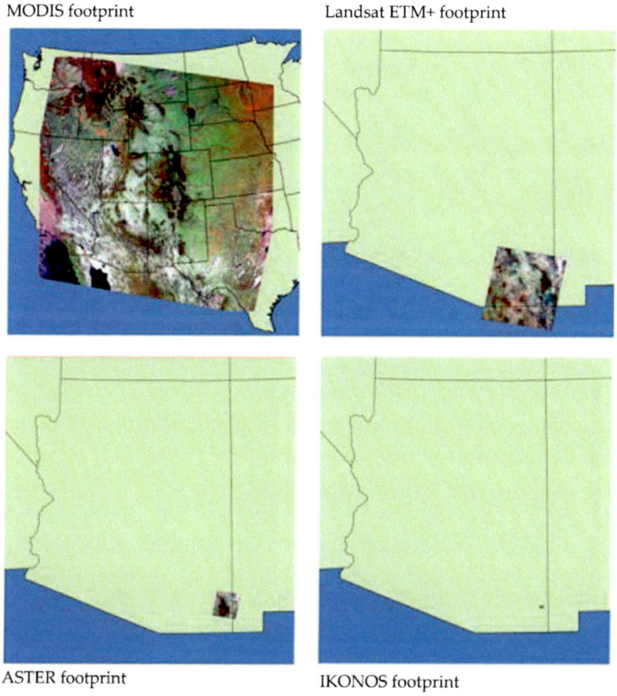

Figure 3.2. *Generally speaking, a coarser sensor resolution results in a larger footprint.*

Source: GeoEye's IKONOS Satellite; Creative Commons License.

Note: The images have different ground widths: MODIS covers 2330 km, the Landsat ETM+ covers 185 km, ASTER covers 60 km, and IKONOS covers 11 km. At this scale, the footprint of the IKONOS image footprint is a mere speck.

Spectral characteristics that influence an image's resolving power involve the number of bands, band placement, and bandwidth. Next, the information regarding the EM spectrum. The spectral bandwidth refers to a range of wavelengths detected in a specific image band along the electromagnetic spectrum. Many multi-spectral sensors have wider bands to maximize light detection and ensure a strong signal-to-noise ratio. However, in certain cases, narrower bands may be more desirable. For instance, by utilizing specific wavelengths, it is possible to identify subtle variations among similar vegetation (Martinelli et al., 2015) (Figure 3.3).

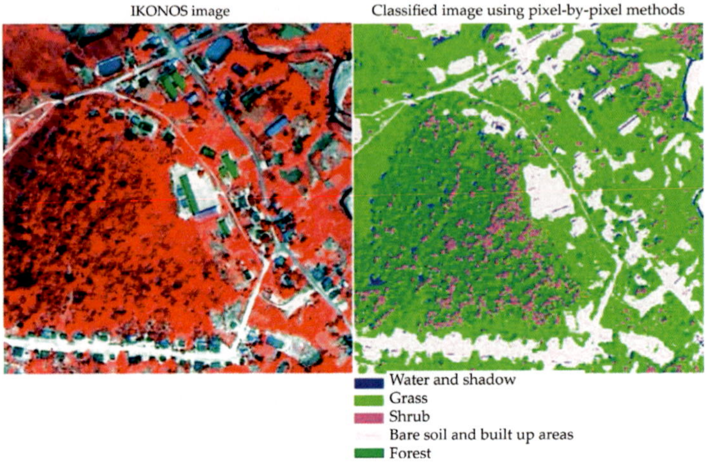

Figure 3.3. *When employing traditional pixel-□y-pixel classification methods to generate maps from high-resolution remotely sensed data, issues may arise since the feature itself is not categorized, but rather its individual components that are considerably larger than the spatial resolution.*

Source: GeoEye's IKONOS Satellite; Creative Commons License.

The part of the electromagnetic range that is used for a certain image band is set by band placement. For instance, one band may identify thermal wavelengths whereas another band may identify blue wavelengths. The specific characteristics of the features under investigation determine the significance of certain bands. For instance, the near-infrared and red bands are helpful for mapping vegetation because green vegetation reflects most NIR light, and absorbs most of the red light (Mulla, 2013).

Note: A land cover map was created using the maximum likelihood method by classifying the IKONOS image on the left. This image has a spatial resolution of 1 m and covers a 400 × 400 m area (ID PO_91586, April 29, 2002). Smooth homogeneous surfaces like bare soil, roads, and rooftops are classified accurately. Some characteristics, meanwhile, are incorrectly categorized as a collection of several classes. For instance, the woodland on the left-hand side of the picture was identified as a combination of grass, shrub, shade, and forest. While considering the forest components, it is often unnecessary to include such detailed information when creating a land cover map. There are ongoing efforts to develop new algorithms to tackle this issue.

The next spectral variable refers to the number of bands a sensor can capture, also known as "spectral resolution." In visual interpretation, this aspect is typically of lesser significance as the focus is on employing at one time, one or three bands. However, it can be crucial to employ automated classification methods. The process of classification of an image entails the translation of pixel values within a satellite image to categories that hold significance. Put simply, every pixel in the image gets a particular label, like "forest" and/or "non-forest." The image products that contain a large number of bands, typically exceeding 100, are referred to as "hyperspectral images." Understanding the significance of each band is crucial as they serve as sample points that record the intensity of light for specific wavelengths. By plotting these points, a spectral curve can be generated. Having a greater number of bands will result in a more seamless curve when the values or radiance are graphed for every band available at a specific pixel (Bauer et al., 2014).

Understanding the spectral curves helps understand the image by identifying various features. Figure 3.4 shows this idea by contrasting two spectral curves: one is from a Landsat ETM image with seven bands, including six in the NIR and visible ranges, and the remaining has 242 bands from a Hyperion image. Hyperion serves as a test system to determine the necessity of hyperspectral sensors for potential Earth remote sensing satellites. Theoretically, minor variations in vegetation kinds can be detected using this incredibly detailed spectral data. For instance, spectranomics, or the use of aerial hyperspectral sensors to monitor the chemical characteristics of the tree canopies—can be used to identify the diversity of trees in moist tropical forests (Asner & Martin, 2008). Working in hyperspectral RS often requires the utilization of hyperspectral sensors (airborne) that offer a significant level of spatial resolution.

The *sensitivity of the sensor* is determined by a dynamic range and quantization, which refers to the digital numbers that may symbolize the range extremes. The value range allowed for a specific data type defines the level of detail that the sensor can detect within this range. There are certain limits that sensors have, where they are unable to register a signal below a certain threshold and become saturated when the radiance increases beyond a certain point. For instance, the values of data from a Landsat TM can vary from 0 to 255, while the values of IKONOS range between 0 and 2048. The potential value range is commonly called the sensor's radiometric resolution. Certain sensors offer the option to adjust the gain settings, allowing for the capture of both low and high radiance values within a particular bandwidth by using two separate bands instead of just one (Kempeneers et al., 2011).

The *repeat frequency*, or temporal resolution, refers to the minimum time re□uired for a specific feature captured by remote sensing to be recorded twice.

For satellites, this is decided by the acquisition schedule and the orbit track. The frequency at which the sensor captures images of a region can have a significant impact on its ability to detect and analyze changes over time, such as the growth and loss of deciduous vegetation, as well as document events like floods (Bruzzone & Bovolo, 2012). If enough images have been collected over a year for a specific area, a temporal profile can be generated (Figure 3.5) by utilizing the individual image bands or vegetation indices. Temporal profiles or signatures have proven effective in distinguishing various vegetation types (Defries & Townshend, 1994). With an option of a high repeat frequency, there is an increase in the likelihood of obtaining imagery with no important cloud cover which is crucial for regions with constant cloud cover.

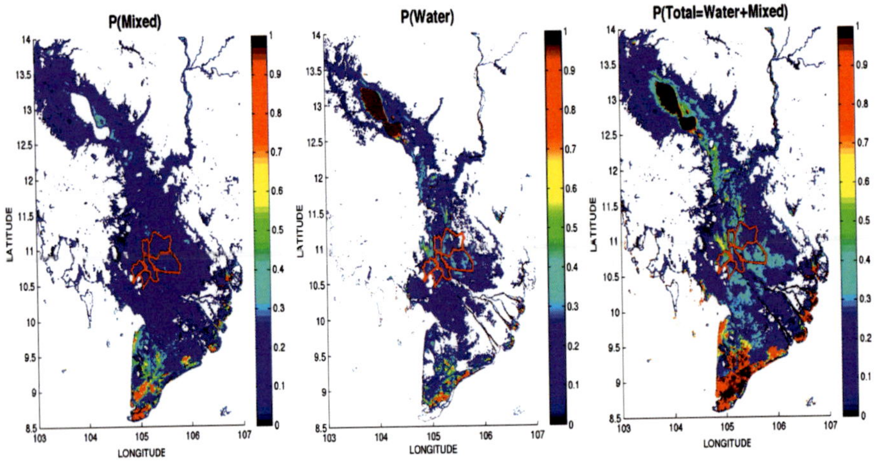

Figure 3.4. *Pro□a□ility of mixed (left); water (middle); and total = mixed + water (right) pixels, over the 2001–2017 period, over the Lower Mekong Basin. The six zones considered in this study over the Cambodia-Vietnam frontier.*

Source: Aires, F., Venot, J.-P., Massuel, S., Gratiot, N., Pham-Duc, B., & Prigent, C. (2020). Surface water evolution (2001–2017) at the Cambodia/Vietnam border in the Upper Mekong Delta using satellite MODIS observations. Remote Sensing, 12(5), 800. https://doi.org/10.3390/rs12050800.

A satellite's orbit controls its height above the surface of the Earth as well as how often it crosses the same location on Earth (Wertz & Larson, 1999). Determined by the imagery acquisition frequency, the geostationary satellites can have a shorter repeat frequency, as they stay in sync with the Earth's rotation and maintain a fixed position relative to the Earth's features. Geosynchronous orbit satellites are positioned 35,786 kilometers above the surface of the Earth, immediately over the equator. Sensors on these satellites usually capture imagery with a lower spatial resolution, typically at the kilometer scale. An example of such imagery is the one produced by the US National Oceanic and Atmospheric Administration Geostationary Operational Environmental Satellite (GOES) program (Hirschmugl et al., 2007).

Earth RS satellites are typically placed in the polar sun-synchronous orbit. As they fly above a particular area, the period of the day remains the same, while each pass over the equator occurs at a different longitude. For instance, the Terra and Aqua pair of missions is set up such that the Terra satellite observes the same region at 10:30 AM–10:45 AM local time, while the Aqua satellite traverses the equator at approximately 1:30 PM local time every day. Typically, Earth remote sensing missions utilize sun-synchronous orbits that are situated at an elevation of several 100 kilometers. Valuable data can be obtained from manned spacecraft like the International Space Station (ISS) and the NASA Space Shuttle, which remain in low Earth orbit and capture images at various stages of their missions (Liao et al., 2014).

Certain sensors in polar orbit can capture several images of the same region within a single day, such as MODIS. Certain sensors can be directed in various directions, enabling them to capture images from different angles (e.g., SPOT). This capability significantly decreases the time it takes to capture data on a specific feature since the sensor has no limits to capturing images directly below the satellite, known as nadir orientation.

The frequency at which a sensor can capture images determines how often a specific area can be observed. However, it should be noted that data ac□uisition is not always guaranteed. Satellite sensors are typically designed to capture specific images. For instance, despite the collection of Landsat data over the years, it was only in 2001 with the implementation of the Landsat 7 long-term ac□uisition plan, that the satellite was directed to capture images of the Pacific coral reef areas (Arvidson et al., 2001).

The data acquisition capacity of a sensor is constrained by factors such as the receiving station's network to download imagery, memory storage capacity, and the available energy to power the sensor (Ayad, 2005).

Image acquisition timing is another important factor to consider in relation to repeat frequency. There are usually one or more ideal times to photograph characteristics that vary over time, like deciduous trees, and events, like floods. For instance, acquiring images during both leaf-off and leaf-on phases can greatly assist in identifying deciduous vegetation. The timing of certain imagery features is a crucial factor to consider when interpreting them (Kangas et al., 2018).

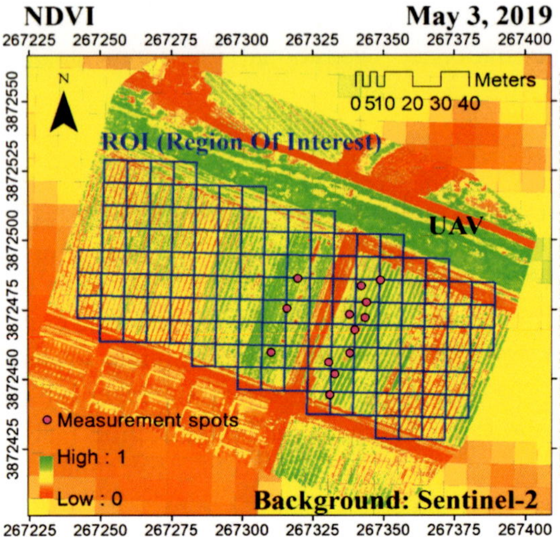

Figure 3.5. *Different forms of vegetation can be characterized by using vegetation indices temporal profiles, like the normalized difference vegetation index (NDVI).*

Source: Ryu, J.-H., Na, S.-I., & Cho, J. (2020). Inter-comparison of normalized difference vegetation index measured from different footprint sizes in cropland. Remote Sensing, 12(18), 2980. https://doi.org/10.3390/rs12182980 (Creative Commons License).

Note: At the top, the four images showcase NDVI values obtained from advanced very high-resolution radiometer (AVHRR) information for four distinct seasons in Türkiye and its adjoining regions. The images show that areas with dark green have the greatest NDVI values, while areas with tan have the lowest. It also displays the temporal profile for the region beneath the three dots marked on the top-left NDVI image. Every dot on the map represents a distinct type of vegetation cover, including sparsely vegetated steppe, evergreen forest, and deciduous broadleaf forest. The plot was made using NDVI measurements from 24 biweekly composites during the year 2000.

3.2.2. Working at Different Scales: The Conceptual and Practical Limitations

While remote sensing provides a wide range of techniques and data sets for working at various scales, it is important to be aware of certain practical limitations before deciding to utilize remote sensing. Typically, these factors are connected to the resources provided by the institution or project for working with remote sensing data, as well as the geographical location of the investigation area (Collier et al., 2012).

From the very beginning stages of a project, it is important to assess the potential benefits of incorporating remote sensing. If remote sensing can significantly contribute to the project, it is crucial to evaluate the resources and capabilities available to work with this type of information. After deciding to integrate remote sensing into the project, the potential limitations of price and imagery availability can often hinder the extent of what can be achieved. While image search capabilities continue to improve, finding suitable imagery can still be a time-consuming task. Additionally, there may be instances where the desired data is not yet available (Treiman, 1975).

There is a vast collection of aerial and satellite photography that is accessible to the public, although it may not always meet the requirements of a particular project. Many of the available satellite images have a spatial resolution that is medium or coarse, typically greater than 15 meters. However, these images are often restricted to the data stored in an archive. If someone wants high-resolution images, they will probably need to buy the data. There are instances where the cost can reach up to US \$100/km^2 for precision-corrected on-demand (processed to precisely match a map base) high-resolution imagery (Paim et al., 2008).

Finding suitable imagery can be quite challenging, particularly in regions with persistent cloud cover like many tropical areas. Locating imagery can pose a challenge when timing is crucial, such as for studying phenology, seasonal flooding, or distinguishing between evergreen and deciduous vegetation. Within certain archives, like the global Landsat GeoCover archive (Tucker et al., 2004), it is not uncommon to find only a single image available over 10 years. Starting from the end of 2008, the US Geological Survey (USGS) began distributing digital imagery from the Landsat archive at no cost.

It is sometimes challenging to locate the desired information when searching through archived data. Even though the data previously existed, there is no assurance of their current availability. Locating and accessing older historical data can be quite challenging and costly, especially when it involves searching through archives that are not easily searchable or in a digital format. Typically, individuals are required to make a payment for such services. Many valuable data sets have been and are still being lost because of the deterioration of the digital or photographic media they are stored on (Faundeen et al., 2004).

There may be conceptual issues related to scale that can restrict the use of remotely sensed data for the specific problem people are working on, especially when it includes local human communities. For example, Walker & Peters (2007) used aerial photographs taken between 1965 and 1995 to examine how the forest had changed in two districts of Malawi. Although the photographs provided clear evidence of forest loss over time, the local communities failed to see the same extent of loss. This difference arose because community members concentrated on smaller areas with trees which were likely missed when analyzing the aerial photographs (Sorooshian et al., 2011).

3.2.3. Using Remote Sensing to Study Social Phenomena

Working with social scientists is necessary for addressing problems pertaining to social phenomena. The social science literature contains numerous instances of projects that have utilized remotely sensed data (Dennis et al., 2005). In the past, remote sensing analysts have frequently performed economic and social analyses without seeking input from social scientists (Turner, 2003). Using traditional remote sensing methods may overlook crucial causal links between the social, political, and natural realms when assessing the effects of humans on the environment, or vice versa. For instance, data from remotely sensed sources can only be used to infer, not conclusively obtain, information regarding land tenure.

To acquire information on property regimes, including land tenure, it is essential to consult with a community member or a social scientist who is well-versed in the institutional arrangements of your study area. When monitoring changes over time, the temporal information gathered by the remotely sensed data may not align with the exact time frame of the political and social developments (Jensen & Cowen, 1999). For example, one may not have remotely sensed data obtained during periods that align with shifts in governance or natural resource policy. There is also a big disparity between human perception of the spatial information and space that remotely sensed data provides, as was described in the preceding section. In summary, numerous features and events that remote sensing analysts may not detect are of utmost importance for research in social science. While this book does not delve into the social sciences with remote sensing, it is worth considering collaborating with social scientists to better understand the interactions between humans and their environment (Gatrell & Jensen, 2008).

3.3. ACCURACY

Remote sensing methods are commonly employed to transform raw or partially processed satellite or aerial imagery into a format suitable for computer models geographic information systems (GIS) or computer models. Having a deep understanding of and accurately measuring the reliability of such products is a crucial aspect of this process (Johansson et al., 2008). An overview of the significance of accuracy evaluation and the interpretation of accuracy data will be given in this section. There will not be a discussion of the specifics of creating and implementing an accurate assessment. For more in-depth information, people can find a selection of recommended reading materials online (Sawaya et al., 2003).

3.3.1. Types of Error

Accuracy can be categorized as either thematic or spatial. The degree to which a derived product matches a reference map or other piece of data that allows precise location data—such as latitude and longitude coordinates—is referred to as spatial accuracy. This is closely tied to the spatial characteristics of the data used to create the final product (Wang & Ellis, 2005). When utilizing satellite imagery, the precision of the final product depends on the satellite image accuracy employed. Errors in spatial measurements can be expressed in different ways. One common method is to use the root mean square error (RMSE) distance, which indicates the average distance from the true location, such as +15 m. Another approach is to express the error using a proportion of the dimensions of the pixels, like plus/minus half the pixel (Lieberman & Cunningham, 2009).

Thematic accuracy refers to the level of agreement between the classes depicted on a map and the actual features present on the ground. For instance, when the thematic map suggests the presence of forest at a specific location, is there indeed forest on the site? Mistakes are bound to happen during the classification process, and deciding on a reasonable amount of error demands careful consideration. Several factors can influence the accuracy of thematic classification. These factors include the type and quality of the base data, the expertise of the analyst, the processing methods used, the analyst's knowledge about the area under classification, and the level of detail in the classification scheme.

When dealing with remote sensing imagery, it is advisable to minimize the number of processing steps required to achieve the final result. In the processing chain, every step involves some data alterations that, when combined, can have a significant impact (Damaso et al., 2020).

3.3.2. Defining Acceptable Accuracy

Since products created from remotely sensed data inherently contain spatial and thematic errors, the objective is to create a product with a reasonable level of errors. The concept of "acceptable" is frequently not determined by what is best for the project but by the resources at hand. Estimating the level of accuracy achievable with specific methods for particular research can be challenging, but discussions with peers and drawing from experience can provide valuable insights. It can be helpful to read remote sensing journals, although it is worth noting that they often focus on the strengths of specific methods and may not provide a comprehensive discussion of their limitations, such as the cost and time involved in producing the final product (Olofsson et al., 2014).

When making a theme map, one rule of thumb is to utilize as few classes as possible because the individual class's accuracy will decline as the quantity of thematic classes increases. After reviewing numerous journal articles on forest classifications, Ross Nelson (2002) found some interesting insights regarding the accuracy of land cover categorizations using Landsat Thematic Mapper data:

- High 90% accuracy rates for soil/vegetated, water/no water, and forest/non-forest categories.
- Conifer/hardwood: Approximately 80–90%.
- Genus: Approximately 60–70%.
- Species: Between 40 and 60%.

Based on the results, it seems that extracting too much thematic detail from the satellite imagery can negatively impact our accuracy per class. Considering this, it would be advisable to choose the most practical minimum number of classes (Kim et al., 2017).

3.3.3. Measure and Report Accuracy

It can take a lot of time and money to do a good accuracy assessment of any product made using remotely sensed data. This is the main reason why this important step is not always done in a mapping project. Often, accuracy assessment is initially incorporated into the project design but then dropped or significantly altered to prioritize time or cost savings.

When faced with such situations, it is possible to make an informed estimation of the accuracy by considering the data employed, the analyst experience, and the methods used. However, ultimately, it remains a mere estimation (Swets, 1988).

Accuracy assessment methods are derived from the context of statistical sampling. When assessing accuracy, it is important to compare the results obtained from remote sensing analysis, such as land cover maps, elevation maps, and timber volume maps, with the real-world features noticed in the field. For instance, using a land cover map, one would compare various points on the map to their e□uivalent on-field points to create a statistical representation of their level of similarity (Eusebi, 2013). Assessing accuracy involves four essential steps:

- Create a sample design;
- Gather validation data;
- Compile validation data; and
- Analyze the obtained data.

Having a sample design is crucial for minimizing bias. While developing the sample design, it is important to address the following questions:

i. How many samples should be selected?
ii. How are the samples going to be distributed?
iii. How is a sample region going to be determined?
iv. How is the field sampling going to be carried out?

The quantity and geographic distribution of samples used will vary depending on the chosen sampling framework. There are formulas available for calculating the optimal number of sample points per class, but in reality, the amount of sample points per class is often determined by constraints such as time and budget. It is generally recommended to have around 50 sample points per class, according to Congalton and Green, although this may not always be feasible. To make sure that the sampling design is strong enough to produce accurate statistics, either consult with a specialist in RS accuracy assessment or follow the instructions found in a book or journal article (Makridakis, 1993).

If the data are in a vector format, the sample unit or area can be made up of polygons, individual pixels, or clusters of homogenous pixels. It is not advisable to use individual pixels due to the challenges in accurately determining their boundaries on the ground. When working with groups of similar pixels, the size is influenced by the image resolution to some extent. There are no specific guidelines for this, but images with higher resolution generally have a greater number of pixels in each block compared to images with lower resolution. It appears that a 3×3-pixel block is commonly used with Landsat imagery (Bowring et al., 2012). Choosing the method for collecting validation data from the field is the final □uestion. There are two commonly used methods: conducting field visits and collecting samples from video footage or aerial photographs. Regardless of the approach taken, it is crucial to precisely locate this data on the image. The use of GPS receivers has greatly simplified this process. Using a digital camera with GPS capabilities and a compass is a widely used technique for capturing field data. This method allows for precise recording of the location where each photograph was snapped and the direction in which the camera was pointed (Warner et al., 2012). It is important to use the same class labels for the validation data as those used for the map data when collecting validation data. It is crucial to ensure that the data obtained for validation is distinct from the data used for classification purposes. Put simply, data that was utilized to enhance the categorization cannot be employed to evaluate the categorization accuracy. A person can guarantee the separation of the two data sets by reserving a random portion of field data for subse□uent accuracy evaluations. It is important to consider the time lag between the map creation using the data and its collection. When there is a considerable time gap between the acquisition of the image used for mapping and the collection of validation data, image interpretation may be correct, but the feature may have changed. It is challenging to create a single accuracy assessment technique that is effective in every circumstance. Numerous factors can vary significantly from one project to the next. Often, practical constraints like limited access to the field sample points and resource constraints such as time and budget may require adjustments to be made to the sampling design, even if it is theoretically ideal. Regardless of the chosen method, it is crucial to thoroughly document these techniques. This ensures that the user of the finished map is informed about the evaluation process of the accuracy (Lin et al., 2012). Contingency tables, also known as "error matrices," are valuable tools for compiling and presenting accurate statistics. By comparing the mapped data with the field data for the samples, a contingency table is used to assess the agreement. Once the data has been tabulated, accuracy statistics can be calculated using straightforward analysis methods (Table 3.1).

Table 3.1. *Measurement Information for Ground-, Unmanned Aerial Vehicle (UAV-), and Satellite-based Remote Sensing Techniques*

Number	**Measurement Date**	**Ground**	**Airborne**	**Satellite**	**Time Difference**[a]
		Leaf- and Field-Spectrometer	**UAV**	**Sentinel-2**	
1	March 28, 2019 March 29, 2019	O –	O[b] –	– O	–1 day +1 day
2	April 16, 2019 April 18, 2019	O –	O[c] –	– O	–2 days +2 days
3	May 3, 2019	O	O[d]	O	0 day
4	May 13, 2019 May 17, 2019	– O	– O[e]	O –	–4 days +4 days

Source: Ryu, J.-H., Na, S.-I., & Cho, J. (2020). Inter-comparison of normalized difference vegetation index measured from different footprint sizes in cropland. Remote Sensing, 12(18), 2980. https://doi.org/10.3390/rs12182980.

[a] *Time difference between satellite and ground/airborne measurements. Measurement time (hh:mm) of UAV –* [b] *12:41–12:50;* [c] *14:14–14:25;* [d] *11:52–12:00; and* [e] *12:13–12:22.*

Accuracy assessments typically provide the overall accuracy of the map, as well as the accuracy for each category within the map. For example, the water class on a land cover map might be highly accurate, while other plant classes may be less precise. The map's utility can be significantly impacted by this, sometimes indicating the need to merge classes. If two forest types, such as coniferous and deciduous, have low accuracy, it might be beneficial to merge them into a single forest class with a higher level of accuracy (Kitchenham et al., 2001).

3.3.4. Understanding the Accuracy Statistics

Examining accurate statistics can be misleading if one does not understand their underlying meanings. Here are some examples of how accuracy is depicted, along with descriptions and comparisons. Let us start with a basic hypothetical example.

As shown in Figure 3.6, the map failed to accurately represent the true landscape, erroneously labeling the entire area as forest when, in reality, 10% of it is non-forest. It is evident that the overall map accuracy is 90%. However, it can be confidently stated that the accuracy for the forest class is 100%, as the entire forest area was accurately classified as such. There are various methods to depict accuracy, as demonstrated by the information provided. The contingency table illustrates two types of accuracy: producer's accuracy and user's accuracy (also known as consumer's accuracy). Reporting all three types of accuracy—overall, producer's, and user's—is considered best practice. However, it is common for reports in published literature to only provide overall accuracy, which can be misleading. User's accuracy is particularly crucial for decision-making and management (Helmus & Babchishin, 2017).

Producer's accuracy refers to the likelihood of correctly classifying a reference sample on the map. Simply put, if one has firsthand knowledge that a specific area is hardwood (after conducting on-site inspections), what are the chances that the digital map will accurately classify the pixels as hardwood?

Consumer accuracy refers to the likelihood that the map sample corresponds to the actual features on the ground (Nolan et al., 2015). For instance, if one selects a hardwood pixel on the classified map, what are the chances of finding a hardwood stand during a visit to that location in the field?

Accuracy terms commonly used in this field include:

1. **Overall Accuracy:** This term refers to the probability that any given pixel is correctly classified in the image. It is calculated by dividing the number of pixels correctly classified by the total number of pixels.

2. **Producer's Accuracy:** This term refers to the probability of correctly classifying reference data of a specific type. It is calculated by dividing the number of pixels accurately classified as forests by the total number of ground reference pixels in the forest.
3. **Omission Error:** This term refers to the likelihood of mistakenly excluding a pixel from a class. It is calculated by subtracting the producer's accuracy from 1.
4. **Consumer's Accuracy:** This term refers to the probability that a point on the map accurately represents what is on the ground. It is calculated by dividing the number of pixels accurately classified as forests by the total number of pixels categorized as forests (Höhle & Höhle, 2009).
5. **Commission Error:** This term refers to the likelihood that a pixel would be included in a class when it should have been omitted. Commission error is calculated as 1 minus the consumer's accuracy.

Other terms associated with accuracy assessments include:

1. **Average (Class) Accuracy:** Calculating the average accuracy by considering the accuracies of each individual class.
2. **Average Accuracy:** Calculating the average accuracy involves adding up the accuracy levels for each class and then dividing that sum by the total number of classes.

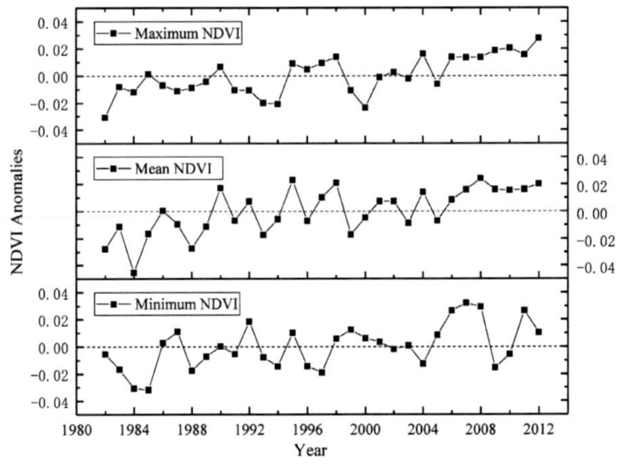

Figure 3.6. *Anomalies of minimum, mean, and maximum NDVIs in the MLR-YR from 1982 to 2012.*

Source: Pei, F., Zhou, Y., & Xia, Y. (2021). Application of normalized difference vegetation index (NDVI) for the detection of extreme precipitation change. Forests, 12(5), 594. https://doi.org/10.3390/f12050594 (Creative Commons License).

3. **Kappa:** It is an accuracy statistic that allows for the comparison of at least two contingency matrices. The statistic accounts for chance agreement when adjusting overall accuracy. To statistically test if two contingency tables agree, use kappa (Schober et al., 2021).

3.3.5. Accuracy Assessment Trends

Using a contingency table, as mentioned earlier, is a widely used method for assessing classification accuracy worldwide. However, it does not provide detailed information about the nature of the errors. To address these limitations, researchers are actively developing and implementing alternative methods (Morales et al., 2019). Two key research areas in accuracy assessment currently being discussed in the remote sensing literature are:

i. How to account for the correlation and geographical distribution of errors?

ii. How to quantify and assess the accuracy of measurements using approaches such as fuzzy set theory and probability, which consider the magnitude of the error, rather than simply determining if a class is correct or incorrect?

In the past, accuracy statistics primarily focused on class and overall map accuracy without considering the errors' spatial characteristics. For instance, in certain map regions, accuracy levels may be quite high, while in others, they could be considerably lower. Several factors can contribute to this phenomenon, such as hazy conditions in certain areas of an image or variations in terrain that affect solar illumination. Understanding how errors are distributed spatially throughout the image can be valuable in assessing a map's effectiveness. One method involves visually representing the precise locations of the reference samples used in map validation, indicating their accuracy through the use of colors or symbols. By visually inspecting the map, one can identify if errors are concentrated in specific regions of the image. Research has been conducted to assess the accuracy of different types of land covers (Tran et al., 2005).

The spatial correlation of error is another important factor to consider when analyzing the spatial distribution. As Congalton (1993) pointed out, the chances of misclassifying a pixel are higher when its neighboring pixel is also misclassified. Furthermore, errors in classification fre□uently arise at the boundaries separating various categories, such as distinct land cover types. This is because land cover often transitions gradually, resulting in the blending of reflected energy from multiple land cover types within a single pixel. This error exemplifies the challenge of defining a clear boundary between types of land covers that transition gradually from one to another. This phenomenon is frequently observed in ecotones, which serve as transitional areas between two distinct ecological communities. An effective strategy to address the mixed pixel issue involves representing land cover in a continuous dataset rather than categorizing it into distinct classes (Olofsson et al., 2013).

Evaluating the extent of error is another aspect of research into accuracy assessment research. By analyzing a contingency table, one can determine the accuracy of point or area classification. The said approach fails to account for the varying degrees of error severity or the possibility of multiple cover types coexisting in a given area. For instance, if an area is completely covered by pine woods, it is evident that it should be identified as such. However, if the forest consists of a combination of oak and pine, then either label may be acceptable to some extent. Nonetheless, classifying any of such pixels as water would be entirely inaccurate. By applying fuzzy set theory (Zadeh 1965), one can consider the extent to which the class is correctly identified (its level of class membership) in a specific region. An approach involves assigning a numerical value ranging from 1 to 5 to represent the level of confidence in the classification of different types of land cover within each sample area (Tehrany et al., 2017).

Given the variability in class membership for various types of covers at each sample site, it is possible to calculate the degree of correctness or incorrectness for each sample point and land cover class. While this approach offers a deeper understanding of accuracy, it does demand additional effort in collecting the re□uired field data for validation. This is because the proportion of every type of land cover needs to be recorded, rather than just the primary cover (Fall et al., 2010).

3.4. ASSESS THE INSTITUTIONAL LIMITS OR IN-HOUSE CAPABILITIES

The skill of the analyst or analysts working with the data, as well as the resources (software and hardware) available, are the two main determinants of what may be achieved with remote sensing technology. Careful evaluation of capabilities and available resources can significantly decrease the likelihood of a project being prematurely terminated or conducted without the necessary resources to yield valuable outcomes (Turnpenny et al., 2008).

Determining the most optimal approach for progressing with a project involving RS can pose a significant challenge. This is especially applicable to a small organization that lacks dedicated specialists in remote sensing. This section offers valuable guidance on the evaluation process that organizations can employ to determine when it is appropriate to outsource work and when it is more advantageous to handle it internally. Several factors need to be assessed to determine the most efficient approach for both short- and long-term (Bonazzi

& Antonelli, 2003). Regrettably, there is no universal solution or predefined answer that applies to every circumstance. The following range of questions can be used when determining the best course of action for a remote sensing project. Additionally, delving into the pros and cons of various options can help make an informed decision (Rush et al., 2014).

3.4.1. Whether to Contract Our Work In-House

If the necessary resources are readily available—such as data storage, computer software, image data, and relevant remote sensing analysis experience—it may be advantageous to utilize internal resources for the project. The decision-maker needs to have a clear understanding of the skills and limitations of the company's internal staff. If this understanding is lacking, consider hiring someone to assess the staff's capabilities to determine whether the project requirements can be efficiently met using internal resources (Marvel & Marvel, 2007). If internal resources are unavailable, consider the following options:

i. Educating current employees;
ii. Hiring a worker to finish the job;
iii. A mix of these options.

For many organizations, it can be difficult to justify the cost of hiring a full-time remote sensing specialist. This is often due to a lack of sufficient work to make use of their expertise and experience, as well as the additional software and hardware required. Sharing an expert, either formally or informally, with other groups is one way to find a solution. For instance, larger organizations may have professionals who specialize in remote sensing as part of their team, while smaller organizations may not have this expertise (Stiens & Turley, 2010). These organizations can establish financial or in-kind agreements that encourage collaboration and provide support. Interdepartmental collaborations in academic settings, especially among departments like agriculture, geology, or geography can offer valuable access to remote sensing expertise that may not be readily available within biology departments where conservation biologists and ecologists are typically based (Button & George, 2005).

3.4.2. Training

Understanding the expectations of different training levels can provide valuable insights when considering training options. It is a common occurrence that individuals are sent for training to learn remote sensing methods, but when they come back to their jobs, they often lack the necessary experience to effectively carry out the tasks independently. Several companies provide remote sensing training, often combined with software training, typically lasting 1–2 weeks.

This training equips individuals with the necessary skills to perform common analyses (Salas et al., 2001). Through the right training in remote sensing, one can acquire a wealth of knowledge in a relatively brief time. Nevertheless, becoming well-versed in remote sensing requires a significant commitment to education and learning. When it comes to planning and managing a remote sensing project, experience is invaluable, even if many of the techniques required to utilize this technology are not hard to acquire (Martin et al., 2014).

Basic remote sensing courses typically emphasize the technical aspects of remote sensing, particularly the physics and software algorithms and physics involved, with a specific emphasis on the classification of land cover. Many students who are interested in the practical aspects of RS typically find these courses to be overwhelming and are unsure of where to begin. One way to begin is by prioritizing the understanding of the capabilities and limitations of remotely sensed data. It is important to learn how to obtain relevant imagery for the specific area of interest and how to effectively analyze and interpret that imagery. This section provides a helpful framework for developing these skills (Salas et al., 2012).

Several short-term remote sensing courses focus on training individuals to perform common tasks found in remote sensing projects, like image rectification and classification. These courses are designed to provide practical skills and save time on these simpler tasks. Project managers for remote sensing can be highly successful when they have these qualities and are supervised by an RS specialist (Wiggenhorn, 1990).

There have been several cases where individuals with limited remote sensing training and lacking expert guidance undertook a project, resulting in a product that was ultimately unusable despite significant effort. This underscores the importance of exploring alternatives, such as engaging a consultant or partnering with a seasoned organization to assist with the project's planning and supervision. With a deep understanding of the subject matter, an expert can provide guidance, offer solutions to challenges, and ensure the project stays on course. Meanwhile, less experienced individuals can handle the bulk of the work, allowing them to learn and grow. It is important to emphasize that experience is invaluable, and if sufficient experience is lacking within the organization, it would be wise to bring in experienced individuals for the project (Greenhalgh & Stewart, 1987).

Another aspect to consider when it comes to training courses is the specific focus of the course. There is a wide range of remote sensing courses available, catering to different levels of expertise and specialization. As an example, if someone wants to learn how to use remote sensing to classify land cover, he or she would need a course that covers that as a study area. Certain courses prioritize the utilization of a particular software brand, while others concentrate more on fundamental techniques, assuming that once a person grasps the techniques, he or she can adapt them to various software options (Campbell, 1971). To become an RS expert, one must plan on finishing college-level coursework and gaining significant experience in the desired field. Alternatively, if the goal is to incorporate remote sensing into an existing expertise, it would be more suitable to start with an introductory course. Before enrolling in a short- or long-term course, it is advisable to communicate with instructors or the organizers of the said course to ensure that the course aligns with the expectations (Foucault & Rabinow, 1984).

3.4.3. Finding a Good Remote Sensing Consultant

Consulting organizations can range in size, from small one-person operations to large corporations. Finding the right consultant for the budget and project requirements can be a challenging task, despite the availability of lists of potential consultants online and from various organizations. Gaining insights from the knowledge and expertise of others is a highly valuable way to identify a trustworthy consultant (Richards, 2005). Speaking with and getting input from colleagues who have worked with consultants on similar projects is beneficial. If this is not feasible, ask potential consultants for a list of past clients and communicate with them to determine if their needs were met. Additionally, try to assess the consultant's compatibility with your organization before making a hiring decision (Mumby et al., 1999).

When creating a contract with a consultant, it is important to clearly outline the specifications for the desired deliverables, such as the methods and data archives that would allow for future modifications by another analyst. Additionally, it is advisable to seek assistance from external sources, such as another consultant or a knowledgeable colleague, to help draft the specifications and, equally important, develop a plan to ensure that the requirements are successfully met. Unfortunately, it is common for a project's shortcomings to become apparent months or years after its completion, revealing that the deliverables do not fully meet the intended goals (Green et al., 1996).

Making an informed decision about pursuing an RS project can be accomplished with a little effort and communication with colleagues (Goodenough et al., 1994).

3.5. SUMMARY

After carefully studying Chapter 1 and the remainder of the book, it becomes evident that remote sensing technology presents numerous opportunities.

However, it is crucial to exercise caution and make informed decisions, taking into account the strengths and weaknesses of the data and methods involved. When making choices, it is important to consider the trade-offs between different parameters. These include spatial resolution, file size, cost, image footprint, the ability to mosaic multiple images, and the level of manual or automated processing required. While some general guidelines can assist with these decisions, personal experience and learning from past mistakes are the most reliable sources for making well-informed choices. Additionally, it is crucial to remember that products derived from remote sensing (RS) imagery will inevitably contain some degree of error. However, by measuring and documenting those errors, one can assess their suitability for a specific task.

REFERENCES

1. Arvidson, T., Gasch, J., & Goward, S. N. (2001). Landsat 7's long-term acquisition plan—An innovative approach to building a global imagery archive. *Remote Sensing of Environment, 78*(1–2), 13–26.
2. Asner, G. P., & Martin, R. E. (2008). Spectral and chemical analysis of tropical forests: Scaling from leaf to canopy levels. *Remote Sensing of Environment, 112*(10), 3958–3970.
3. Ayad, Y. M. (2005). Remote sensing and GIS in modeling visual landscape change: A case study of the northwestern arid coast of Egypt. *Landscape and Urban Planning, 73*(4), 307–325.
4. Bauer-Marschallinger, B., Sabel, D., & Wagner, W. (2014). Optimization of global grids for high-resolution remote sensing data. *Computers & Geosciences, 72,* 84–93.
5. Bonazzi, G., & Antonelli, C. (2003). To make or to sell? The case of in-house outsourcing at Fiat Auto. *Organization Studies, 24*(4), 575–594.
6. Bowring, A. L., Peeters, A., Freak-Poli, R., Lim, M. S., Gouillou, M., & Hellard, M. (2012). Measuring the accuracy of self-reported height and weight in a community-based sample of young people. *BMC Medical Research Methodology, 12,* 1–8.
7. Bruzzone, L., & Bovolo, F. (2012). A novel framework for the design of change-detection systems for very high-resolution remote sensing images. *Proceedings of the IEEE, 101*(3), 609–630.
8. Button, M., & George, B. (2005). Why do some reebank es ns prefer in-house to contract security staff? *Crime at Work: Studies in Security and Crime Prevention,* Vol. I, 210–224.
9. Campbell, J. P. (1971). Personnel training and development. *Annual Review of Psychology, 22*(1), 565–602.
10. Carleer, A. P., Debeir, O., & Wolff, E. (2005). Assessment of very high spatial resolution satellite image segmentations. *Photogrammetric Engineering & Remote Sensing, 71*(11), 1285–1294.
11. Collier, D., LaPorte, J., & Seawright, J. (2012). Putting typologies to work: Concept formation, measurement, and analytic rigor. *Political Research Quarterly, 65*(1), 217–232.
12. Congalton, R. G., Stenback, J. M., & Barrett, R. H. (1993). Mapping deer habitat suitability using remote sensing and geographic information systems. *Geocarto International, 8*(3), 23–33.

13. Damaso, K., Williams, P., & Heathcote, A. (2020). Evidence for different types of errors being associated with different types of post-error changes. *Psychonomic Bulletin & Review, 27,* 435–440.

14. DeFries, R. S., & Townshend, J. R. G. (1994). NDVI-derived land cover classifications at a global scale. *International Journal of Remote Sensing, 15*(17), 3567–3586.

15. Dennis, M. L., Scott, C. K., Funk, R., & Foss, M. A. (2005). The duration and correlates of addiction and treatment careers. *Journal of Substance Abuse Treatment, 28*(2), S51–S62.

16. Eusebi, P. (2013). Diagnostic accuracy measures. *Cerebrovascular Diseases, 36*(4), 267–272.

17. Fall, S., Niyogi, D., Gluhovsky, A., Pielke Sr, R. A., Kalnay, E., & Rochon, G. (2010). Impacts of land use land cover on temperature trends over the continental United States: Assessment using the North American Regional Reanalysis. *International Journal of Climatology, 30*(13), 1980–1993.

18. Faundeen, J. L., Williams, D. L., & Greenhagen, C. A. (2004). Landsat yesterday and today: An American vision and an old challenge. *Journal of Map and Geography Libraries, 1*(1), 59–73.

19. Fogassy, E., Nógrádi, M., Kozma, D., Egri, G., Pálovics, E., & Kiss, V. (2006). Optical resolution methods. *Organic & Biomolecular Chemistry, 4*(16), 3011–3030.

20. Foucault, M., & Rabinow, P. (1984). The means of correct training. *Crime: Critical Concepts in Sociology,* 188–205.

21. Gatrell, J. D., & Jensen, R. R. (2008). Sociospatial applications of remote sensing in urban environments. *Geography Compass, 2*(3), 728–743.

22. Goodenough, D. G., Charlebois, D., Matwin, S., & Robson, M. (1994). Automating reuse of software for expert system analysis of remote sensing data. *IEEE Transactions on Geoscience and Remote Sensing, 32*(3), 525–533.

23. Green, E. P., Mumby, P. J., Edwards, A. J., & Clark, C. D. (1996). A review of remote sensing for the assessment and management of tropical coastal resources. *Coastal Management, 24*(1), 1–40.

24. Greenhalgh, C., & Stewart, M. (1987). The effects and determinants of training. *Oxford Bulletin of Economics and Statistics, 49*(2), 171–190.

25. Helmus, L. M., & Babchishin, K. M. (2017). Primer on risk assessment and the statistics used to evaluate its accuracy. *Criminal Justice and Behavior, 44*(1), 8–25.

26. Hirschmugl, M., Ofner, M., Raggam, J., & Schardt, M. (2007). Single tree detection in very high-resolution remote sensing data. *Remote Sensing of Environment, 110*(4), 533–544.
27. Höhle, J., & Höhle, M. (2009). Accuracy assessment of digital elevation models by means of robust statistical methods. *ISPRS Journal of Photogrammetry and Remote Sensing, 64*(4), 398–406.
28. Jensen, J. R., & Cowen, D. C. (1999). Remote sensing of urban/suburban infrastructure and socio-economic attributes. *Photogrammetric Engineering & Remote Sensing, 65,* 611–622.
29. Johansson, M., Lovmar, M., & Ehrenberg, M. (2008). Rate and accuracy of bacterial protein synthesis revisited. *Current Opinion in Microbiology, 11*(2), 141–147.
30. Kangas, A., Astrup, R., Breidenbach, J., Fridman, J., Gobakken, T., Korhonen, K. T., & Olsson, H. (2018). Remote sensing and forest inventories in Nordic countries–a roadmap for the future. *Scandinavian Journal of Forest Research, 33*(4), 397–412.
31. Kempeneers, P., Sedano, F., Seebach, L., Strobl, P., & San-Miguel-Ayanz, J. (2011). Data fusion of different spatial resolution remote sensing images applied to forest-type mapping. *IEEE Transactions on Geoscience and Remote Sensing, 49*(12), 4977–4986.
32. Kim, K. Y., Park, J., & Sohmshetty, R. (2017). Prediction measurement with mean acceptable error for proper inconsistency in noisy weldability prediction data. *Robotics and Computer-Integrated Manufacturing, 43,* 18–29.
33. Kitchenham, B. A., Pickard, L. M., MacDonell, S. G., & Shepperd, M. J. (2001). What accuracy do statistics really measure? *IEE Proceedings-Software, 148*(3), 81–85.
34. Kühlbrandt, W. (2014). The resolution revolution. *Science, 343*(6178), 1443–1444.
35. Liao, A., Chen, L., Chen, J., He, C., Cao, X., Chen, J., & Gong, P. (2014). High-resolution remote sensing mapping of global land water. *Science China Earth Sciences, 57,* 2305–2316.
36. Lieberman, M. D., & Cunningham, W. A. (2009). Type I and Type II error concerns in fMRI research: Re-balancing the scale. *Social Cognitive and Affective Neuroscience, 4*(4), 423–428.
37. Lin, C. J., DeRoo, L. A., Jacobs, S. R., & Sandler, D. P. (2012). Accuracy and reliability of self-reported weight and height in the Sister Study. *Public Health Nutrition, 15*(6), 989–999.

38. Liu, W., & Wu, E. Y. (2005). Comparison of non-linear mixture models: Sub-pixel classification. *Remote Sensing of Environment, 94*(2), 145–154.
39. Makridakis, S. (1993). Accuracy measures: Theoretical and practical concerns. *International Journal of Forecasting, 9*(4), 527–529.
40. Martin, B. O., Kolomitro, K., & Lam, T. C. (2014). Training methods: A review and analysis. *Human Resource Development Review, 13*(1), 11–35.
41. Martinelli, F., Scalenghe, R., Davino, S., Panno, S., Scuderi, G., Ruisi, P., & Dandekar, A. M. (2015). Advanced methods of plant disease detection. A review. *Agronomy for Sustainable Development, 35,* 1–25.
42. Marvel, M. K., & Marvel, H. P. (2007). Outsourcing oversight: A comparison of monitoring for in house and contracted services. *Public Administration Review, 67*(3), 521–530.
43. Meinel, G., & Neubert, M. (2004). A comparison of segmentation programs for high resolution remote sensing data. *International Archives of Photogrammetry and Remote Sensing, 35*(Part B), 1097–1105.
44. Morales-Barquero, L., Lyons, M. B., Phinn, S. R., & Roelfsema, C. M. (2019). Trends in remote sensing accuracy assessment approaches in the context of natural resources. *Remote Sensing, 11*(19), 2305.
45. Mulla, D. J. (2013). Twenty-five years of remote sensing in precision agriculture: Key advances and remaining knowledge gaps. *Biosystems Engineering, 114*(4), 358–371.
46. Mumby, P. J., Green, E. P., Edwards, A. J., & Clark, C. D. (1999). The cost-effectiveness of remote sensing for tropical coastal resources assessment and management. *Journal of Environmental Management, 55*(3), 157–166.
47. Nelson, D. E., Bland, S., Powell-Griner, E., Klein, R., Wells, H. E., Hogelin, G., & Marks, J. S. (2002). State trends in health risk factors and receipt of clinical preventive services among US adults during the 1990s. *JAMA, 287*(20), 2659–2667.
48. Nilsson, H. E. (1995). Remote sensing and image analysis in plant pathology. *Canadian Journal of Plant Pathology, 17*(2), 154–166.
49. Nolan, J. J., Haas, S. M., Turley, E., Stump, J., & LaValle, C. R. (2015). Assessing the "statistical accuracy" of the national incident-based reporting system hate crime data. *American Behavioral Scientist, 59*(12), 1562–1587.
50. Olofsson, P., Foody, G. M., Herold, M., Stehman, S. V., Woodcock, C. E., & Wulder, M. A. (2014). Good practices for estimating area and assessing the accuracy of land change. *Remote Sensing of Environment, 148,* 42–57.

51. Olofsson, P., Foody, G. M., Stehman, S. V., & Woodcock, C. E. (2013). Making better use of accuracy data in land change studies: Estimating accuracy and area and quantifying uncertainty using stratified estimation. *Remote Sensing of Environment, 129,* 122–131.

52. Paim, R., Caulliraux, H. M., & Cardoso, R. (2008). Process management tasks: A conceptual and practical view. *Business Process Management Journal, 14*(5), 694–723.

53. Pinter Jr, P. J., Hatfield, J. L., Schepers, J. S., Barnes, E. M., Moran, M. S., Daughtry, C. S., & Upchurch, D. R. (2003). Remote sensing for crop management. *Photogrammetric Engineering & Remote Sensing, 69*(6), 647–664.

54. Qin, Z., & Zhang, M. (2005). Detection of rice sheath blight for in-season disease management using multispectral remote sensing. *International Journal of Applied Earth Observation and Geoinformation, 7*(2), 115–128.

55. Richards, J. A. (2005). Analysis of remotely sensed data: The formative decades and the future. *IEEE Transactions on Geoscience and Remote Sensing, 43*(3), 422–432.

56. Rush, H., Bessant, J., Hobday, M., Hanrahan, E., & Medeiros, M. Z. (2014). The evolution and use of a policy and research tool: Assessing the technological capabilities of firms. *Technology Analysis & Strategic Management, 26*(3), 353–365.

57. Salas, E., & Cannon-Bowers, J. A. (2001). The science of training: A decade of progress. *Annual Review of Psychology, 52*(1), 471–499.

58. Salas, E., Tannenbaum, S. I., Kraiger, K., & Smith-Jentsch, K. A. (2012). The science of training and development in organizations: What matters in practice. *Psychological Science in the Public Interest, 13*(2), 74–101.

59. Salin, P. A. (1992). Proprietary aspects of commercial remote-sensing imagery. *Northwestern Journal of International Law & Business, 13,* 349.

60. Sawaya, K. E., Olmanson, L. G., Heinert, N. J., Brezonik, P. L., & Bauer, M. E. (2003). Extending satellite remote sensing to local scales: Land and water resource monitoring using high-resolution imagery. *Remote Sensing of Environment, 88*(1–2), 144–156.

61. Schober, P., Mascha, E. J., & Vetter, T. R. (2021). Statistics from A (agreement) to Z (z score): A guide to interpreting common measures of association, agreement, diagnostic accuracy, effect size, heterogeneity, and reliability in medical research. *Anesthesia & Analgesia, 133*(6), 1633–1641.

62. Sheldrick, G. M., & Schneider, T. R. (1997). SHELXL: High-resolution refinement. In *Methods in Enzymology* (Vol. 277, pp. 319–343).

63. Sorooshian, S., AghaKouchak, A., Arkin, P., Eylander, J., Foufoula-Georgiou, E., Harmon, R., & Hendrickx, J. M. H. (2011). Advanced concepts on remote sensing of precipitation at multiple scales. *Bulletin of the American Meteorological Society, 92*(10), 1353–1357.

64. Stiens, K. P., & Turley, S. L. (2010). Uncontracting: The move back to performing in-house. *AFL Review, 65,* 145.

65. Swets, J. A. (1988). Measuring the accuracy of diagnostic systems. *Science, 240*(4857), 1285–1293.

66. Tehrany, M. S., Kumar, L., & Drielsma, M. J. (2017). Review of native vegetation condition assessment concepts, methods, and future trends. *Journal for Nature Conservation, 40,* 12–23.

67. Tran, H. T., Wang, X., & Pappu, R. V. (2005). Reconciling observations of se□uence-specific conformational propensities with the generic polymeric behavior of denatured proteins. *Biochemistry, 44*(34), 11369–11380.

68. Treiman, D. J. (1975). Problems of concept and measurement in thc comparative study of occupational mobility. *Social Science Research, 4*(3), 183–230.

69. Tucker, C. J., Grant, D. M., & Dykstra, J. D. (2004). NASA's global orthorectified Landsat data set. *Photogrammetric Engineering & Remote Sensing, 70*(3), 313–322.

70. Turner, M. W. (2003). The role of mannose-binding lectin in health and disease. *Molecular Immunology, 40*(7), 423–429.

71. Turnpenny, J., Nilsson, M., Russel, D., Jordan, A., Hertin, J., & Nykvist, B. (2008). Why is integrating policy assessment so hard? A comparative analysis of the institutional capacities and constraints. *Journal of Environmental Planning and Management, 51*(6), 759–775.

72. Walker, P. A., & Peters, P. E. (2007). Making sense in time: Remote sensing and the challenges of temporal heterogeneity in social analysis of environmental change—Cases from Malawi. *Human Ecology, 35,* 69–80.

73. Wang, H., & Ellis, E. C. (2005). Image misregistration error in change measurements. *Photogrammetric Engineering & Remote Sensing, 71*(9), 1037–1044.

74. Warner, E. T., Wolin, K. Y., Duncan, D. T., Heil, D. P., Askew, S., & Bennett, G. G. (2012). Differential accuracy of physical activity self-report by body mass index. *American Journal of Health Behavior, 36*(2), 168–178.

75. Watts, A. C., Ambrosia, V. G., & Hinkley, E. A. (2012). Unmanned aircraft systems in remote sensing and scientific research: Classification and considerations of use. *Remote Sensing, 4*(6), 1671–1692.

76. Wertz, J. R., & Larson, W. J. (1999). *Space mission analysis and design,* Section, 11(2), 501–507.

77. Wiggenhorn, W. (1990). Motorola U: When training becomes an education. *Harvard Business Review, 68*(4), 71–83.

78. Xie, Y., Sha, Z., & Yu, M. (2008). Remote sensing imagery in vegetation mapping: A review. *Journal of Plant Ecology, 1*(1), 9–23.

79. Yang, C. (2020). Remote sensing and precision agriculture technologies for crop disease detection and management with a practical application example. *Engineering, 6*(5), 528–532.

80. Zadeh, L. A. (1965). Fuzzy sets. *Information and Control, 8*(3), 338–353.

Chapter 4

REMOTE SENSING FOR INSECT PEST AND DISEASE MANAGEMENT

Contents

4.1. INTRODUCTION

With the advancement of information technologies such as GIS (Geographic Information Systems), GPS (Global Positioning Systems), and RS (Remote Sensing) sensors, a novel method has emerged for detecting and controlling insect pests. Remote sensing technologies rely on the reception and interpretation of data regarding the Earth's surface without physical contact. RS heavily depends on the analysis of electromagnetic (EM) energy to enhance data collection and interpretation. This technology measures and records the emission and reflection of EM radiation from the area under observation using various sensor instruments (Zhang et al., 2019). Examples of such instruments include radar systems, video cameras, electro-mechanical scanners, and digital cameras. These technologies rely on EM energy and how it interacts with ground targets (Yang & Everitt, 2011).

The EM spectrum is divided into both invisible and visible radiation, including X-rays, gamma rays, UV, infrared, radio waves, and microwaves. Naturally, the sun produces EM energy, which travels millions of kilometers through space at the speed of light to reach Earth's atmosphere and surface. Some of this energy is then reflected from the Earth's surface, while some is absorbed and emitted again as thermal energy. EM radiation consists of a magnetic field and an electric field that is perpendicular to each other. Understanding remote sensing requires knowledge of two key elements of EM radiation: wavelength and frequency. The wavelength, denoted by the Greek letter lambda (λ), represents the length of one complete wave cycle. It may be expressed in meters or smaller units such as centimeters (cm), micrometers (μm), or nanometers (nm), and it is calculated as the distance between consecutive wave crests. The fre□uency, denoted by the Greek letter nu (ν), is measured in hertz (Hz) and represents the number of cycles per second (Yang & Everitt, 2011).

Many fields have reported using highly valuable RS applications, such as weather, agriculture, water resources, forest ecosystems, marine sciences, and insect pest management. Such applications are categorized into the following three primary tactics: monitoring and mapping of different earth resources, prediction, and forecasting model(s), followed by decision-making and management based on the collected information. Prediction and forecasting models rely on retrieving information from bio-and geophysical factors, as demonstrated by Oumar & Mutanga (2011). The RS systems today are deployed on satellites and utilize multi-spectral imaging devices that offer a comprehensive view, fast survey capabilities, precise measurements, and coverage of inaccessible areas. These advancements surpass the limitations of traditional monitoring and mapping methods. For instance, satellite-borne RS systems are utilized to quantify water surfaces or characterize vegetation during floods, providing hydrologists and ecologists with quick and precise data on tracking the water cycle along with its impact on the ecosystem (Malinowski et al., 2015). This work explores different aspects of RS systems along with their applications in agriculture, specifically focusing on the management of insect pests and plant diseases (Eigenbrode et al., 2016).

4.2. REMOTE SENSING SYSTEMS

4.2.1. Principals of RS Systems

Nowadays, EM energy can be generated artificially, and there are no essential physical differences between the artificial and natural sources. EM energy may be detected using either wavelength or frequency. Avery and Berlin divided the EM spectrum into different spectral regions (Figure 4.1).

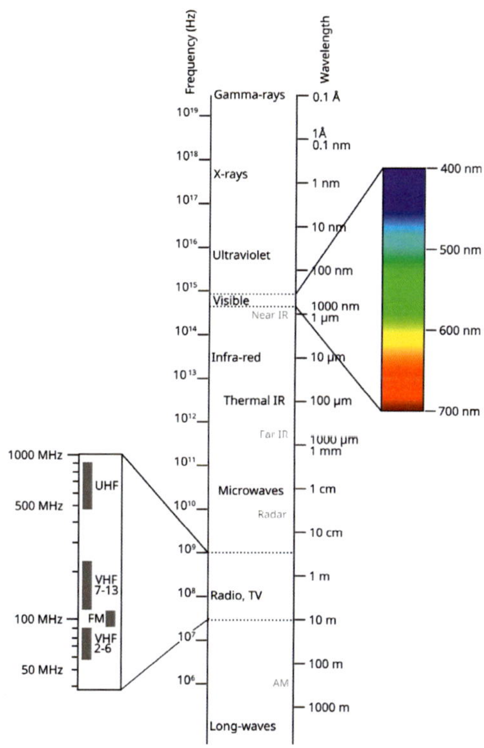

Figure 4.1. *The main spectral regions of the EM spectrum.*

Source: By Original: Penubag. (n.d.). Vector: Victor Blacu. [Own work based on: Electromagnetic-Spectrum.png]. CC BY-SA 3.0. https://commons.wikimedia.org/w/index.php?curid=22428451.

Remote sensors gather data about EM radiation that is reflected or emitted from a distance. Either photographic or electronic equipment can be used to detect EM energy. The first is referred to as a photograph, whereas the second is referred to as an image. Many natural surfaces act like diffuse reflectors, making RS applications relevant for studying the spectral behavior of living plant leaves (Banerjee et al., 2006).

4.2.2. Components of an RS System

RS systems consist of six main components: an energy source, target interaction, sensors, platforms, transmission, and reception processing, and interpretation and analysis processes. These components work together to enable various applications (Freire et al., 2008).

4.2.2.1. Energy Source

It emits EM energy towards the target. There are two main classifications for sensors based on their energy source: active and passive systems:

i. **Passive System:** It measures solar radiation absorbed, emitted (in thermal infrared wavelengths), or reflected (in visible wavelengths) by Earth's materials.

ii. **Active System:** It exemplifies the artificial energy that generates its own EM radiation, including radars, laser fluoro sensors, and microwaves (Lanza, 2005).

4.2.2.2. Interaction with the Target

After the energy passes through the atmosphere, the interaction with its target occurs, depending on the radiation and properties of the target. There are several different interactions that can occur, the substance can absorb the energy, emit it upwards, scatter it, or reflect it (Palomar et al., 2011).

1. **Sensors:** Several devices were set up to collect and monitor the EM radiation scattered or emitted from the target, such as cameras and scanners. The two types of remote sensors are microwave sensors (that operate between 109 and 1012 HZ in the EM wave frequencies) and optical sensors (that operate in the thermal infrared, middle, near, or visible wavelength regions) (Cheng & Chen, 2009).
2. **Platforms:** It includes various types of sensor carriers such as ground-based handheld radiometers, airborne aircraft NASA AVIRIS Sensor, and space shuttles and satellites.

3. **Transmission, Reception, and Processing:** It involves the management of information captured by the sensor, then transmitted to the station on Earth and stored as digital images, photographs, or in other forms (Rogers & Randolph, 2003).
4. **Interpretation and Analysis:** The process of analysis of the collected data and applying various image processing tools (software and hardware) to extract destination information. Afterward, the information obtained is utilized to address any particular issue (Hall et al., 2016).
5. **Applications/Utilization:** It demonstrates the practical implications of the RS technique, showcasing how the collected data can be effectively applied to tackle any particular problem (White et al., 1978).

4.3. PLATFORMS AND DEVICES USED IN RS SYSTEMS

There are two main types of remote sensing (RS) devices: non-imaging devices, such as spectroradiometers, and imaging devices, including various types of cameras. RS can be classified based on different wavelength regions, including microwave, thermal, reflective, and visible infrared RS. Additionally, RS is categorized into three types based on the platforms carrying the sensors: airborne, space-based, and ground-based sensors. Ground-based platforms consist of structures like towers, vehicles, and ground-based equipment, typically reaching heights of up to 50 meters (Zhang & Kovacs, 2012). Examples of ground-based platforms include the Atmospheric Radiation Measurement (ARM) program by the Department of Energy (DOE) and NASA's AERONET network.

Airborne sensor platforms, such as airplanes, helicopters, balloons, and high-altitude aircraft, provide a convenient and flexible means of covering distances ranging from tens to thousands of meters (up to 50 km), with minimal impact from atmospheric effects. However, airborne observations are limited in duration to only a few hours, unlike space-borne sensors. Examples of airborne sensor platforms include research aircraft from organizations like the National Oceanic and Atmospheric Administration (NOAA), the US National Center for Atmospheric Research (NCAR), and NASA.

Space-borne platforms overcome the challenges posed by limited data collection time, atmospheric conditions, and lower altitudes. These platforms include manned space vehicles, satellites, shuttles, and rockets (Nuzzo et al., 2015), operating at altitudes ranging from 100 to 36,000 km. Photographic systems are used to record the orientation of rockets during flight. However, rockets have drawbacks, including limited coverage of fine weather and fast vertical firing, which restricts their cameras' ability to capture images in a short time frame (approximately 8 minutes). In contrast, satellites orbit the Earth along a predetermined path and provide continuous Earth observation from both high-altitude (36,000 km) and low-altitude (700–1500 km) orbits.

There are three types of satellites based on their orbital characteristics: equatorial geosynchronous orbits (e.g., NOAA weather satellites), sun-synchronous orbits, and polar geosynchronous orbits (e.g., NPOESS: National Polar-orbiting Operational Environmental Satellite System). The first international satellite program, overseen by NASA, was launched in 1972 to gather multispectral image data of the Earth's surface using Landsat sensors. In 1982, the French government initiated the SPOT program for satellite sensors. Since then, advancements in high-resolution satellite systems have been made. Satellites like IKONOS, QuickBird, GeoEye-1, and WorldView-2, launched in the late 1990s and early 2000s, now provide high-resolution imagery (Zrelli, 2022). These systems were utilized in various agricultural methods, as documented by Malinowski et al. (2015). The latest space-borne platform is a manned space vehicle allowing for on-spot human control throughout the imaging of Earth from space. In addition, this type of platform allows for the transportation of RS devices into the space and their subsequent return to the ground for recalibration and refurbishment (Ricci, 2010). The latest satellite sensors feature advanced super-spectral imaging systems that can capture images across a wide range of spectral wavelengths, including the visible, near-infrared (NIR), and short-wave to thermal infrared portions of the spectrum. These sensors have narrower bandwidths, allowing them to capture the intricate spectral characteristics of the target (Fernando et al., 2018).

When selecting an RS system, it is important to consider different factors like temporal, radiometric, spectral, and spatial resolutions. The spatial resolution indicates the smallest object that can be observed in the field or identified in an image. A pixel is the basic building block of an image. With a one-meter spatial resolution, each pixel represents an area of 1 m^2. A spectral resolution is indicative of the wavelength width of each band and the number of bands. Higher spectral resolution images can help to identify shorter wavelength widths. Nevertheless, the radiometric resolution plays a crucial role in determining the sensitivity of the remote sensor to varying reflectance levels. Lastly, the temporal resolution refers to the frequency at which the RS platform scans an area. For instance, data is provided by normal orbiting satellites every time they cross a region. Additionally, geostationary satellites are utilized to provide uninterrupted data sensing and scanning across the identified region (Watts et al., 2012).

4.4. UTILIZING RS SYSTEMS FOR AGRICULTURAL APPLICATIONS

Remote sensing (RS) plays a crucial role in ensuring sustainable agricultural management. Several applications have been developed in agriculture, including crop growth monitoring, crop discrimination, soil moisture estimation, stress detection, plant disease infestation monitoring, soil fertility analysis, insect infestation detection, underground water level monitoring, evapotranspiration computation for crops, site-specific management, water salinity measurement, yield prediction estimation, and crop acreage calculation. Precision agriculture helps farmers understand the complex spatial relationships between abiotic and biotic environments. It also shows how effective applications and precise amendments of pesticides, fertilizers, and herbicides can help lower the amount of chemicals used and keep crops profitable (Ricci, 2010). For example, the implementation of site-specific crop management (SSCM) can greatly enhance productivity in crops while minimizing the input requirements in precision agriculture (Peralta et al., 2016). In addition, Fitzgerald (2000) utilized multi-spectral RS to identify the initial presence of the mites in cotton fields by observing color variations and subsequent alterations in the overall appearance of the crop canopy. Field scouts used portable GPS equipment to identify infected areas on the map and verify the population of mites in the areas under surveillance. This allowed for the discrimination of particular cotton field zones that needed pesticides (Zhan et al., 2014).

RS is a valuable tool that provides important information for integrated pest management in various agricultural methods. For instance, different types of aerial RS, such as multi- and hyper-spectral sensors, were employed for agricultural purposes (Figure 4.2). Various systems utilize a video and/or single-band camera containing specialized filter lenses to capture images in specific wavelengths. These wavelengths include the thermal and NIR bands; and red, green, and blue bands. Several authors have researched the use of air-borne RS to detect and analyze the field damage due to insect pests. In addition, it has been extensively utilized in the mapping of crop diseases and evaluating their effects on crop productivity. Several studies have been conducted on this topic, including those by Pretorius et al. (2017); and Dhau et al. (2018). Delineate damage, for instance, is the source of several harmful pests, such as bark beetles and hemlock loopers; color and color-infrared aerial photography using a standard camera has been employed successfully. In their research by Yang et al. (2009), they mapped the damage due to aphid species in the Mississippi River delta. Additionally, they used RS to predict the areas where insects are most likely to attack wheat crops.

4.5. RS APPLICATIONS FOR PEST MANAGEMENT

Having a solid understanding of basic information is crucial for implementing successful pest management programs. Valuable insights can be gained by having access to information on plant diseases, weed infestations, and insect damage levels to make informed decisions about management programs. RS technology is valuable for gathering data that can be used to support decision-making in pest management programs. It helps with monitoring, followed by mapping, and even predicting the outbreak patterns. Efficient and accurate monitoring of insect populations is crucial for enhancing pest control. In addition, the speed and efficiency of forecasting along with the reduction of managerial expenses have been greatly enhanced with computer-based systems. Forecasting for pests is a crucial component of the IPM strategy. Utilizing biophysical methods, early warnings and accurate forecasts provide valuable time for effectively managing pest attacks.

Figure 4.2. *A two-camera multispectral imaging system mounted on an aluminum rack.*

Source: Yang, C., Westbrook, J. K., Suh, C. P.-C., Martin, D. E., Hoffmann, W. C., Lan, Y., Fritz, B. K., & Goolsby, J. A. (2014). An airborne multispectral imaging system based on two consumer-grade cameras for agricultural remote sensing. Remote Sensing, 6(6), 5257–5278. https://doi.org/10.3390/rs6065257 (Creative Commons License).

This proactive approach optimizes pest control and minimizes crop loss, ultimately reducing the overall cost of cultivation (Khater, 2012). Identifying crop stresses such as disease development, pest infestation, and nutrient deficiency, and monitoring drought through RS techni□ues prove to be highly valuable. RS enhances temporal and spatial resolution in pest monitoring, surpassing traditional methods like sex pheromone traps for selective species monitoring, long-term suction traps for tracking insect pest migration patterns, and light traps for multiple species (Prasad & Prabhakar, 2012). All these technologies, rooted in plant protection research against the major pests, play a crucial role in achieving sustainability in agriculture.

RS techniques have been widely explored globally to determine their potential for automating or enhancing this system. RS technologies are utilized to acquire and process multi-spectral images (Andreo, 2013). In addition, the application of RS and other GIS and/or GPS technologies is used to detect and map insect infestations. Aerial videography, GPS, and GIS are used together to make regional maps of insect infestations throughout a big forest and farmland belt. These combinations assist the agronomic advisor in making informed decisions. Morales and Jones conducted an RS technique to study the impact of whitefly infestations on different crops in 2004 (Figure 4.4). The decline in plant vigor brought on by feeding activities is an example of the direct damage, while the sooty mold fungus growth on honeydew secreted by the feeding nymphs and the spread of significant plant diseases are examples of the indirect damage. In a study conducted in Shuibian, China, Yan et al. (2017) implemented a volunteered geographic information (VGI)--based IPM system.

Figure 4.3. *Unmanned aerial vehicle acquiring images over experimental orchard.*

Source: Bulanon, D. M., Lonai, J., Skovgard, H., & Fallahi, E. (2016). Evaluation of different irrigation methods for an apple orchard using an aerial imaging system. ISPRS International Journal of Geo-Information, 5(6), 79. https://doi.org/10.3390/ijgi5060079.

The goal was to evaluate the efficacy of this approach in analyzing infestation levels and predicting the striped rice stem borer outbreak timings. The researchers relied on voluntary cellphone information from farmers and examined the spatiotemporal infestation patterns of the borers that are under investigation. The technique has a distinct advantage which improves integrated pest management (IPM) for the target pest (Dong et al., 2021) (Figure 4.3).

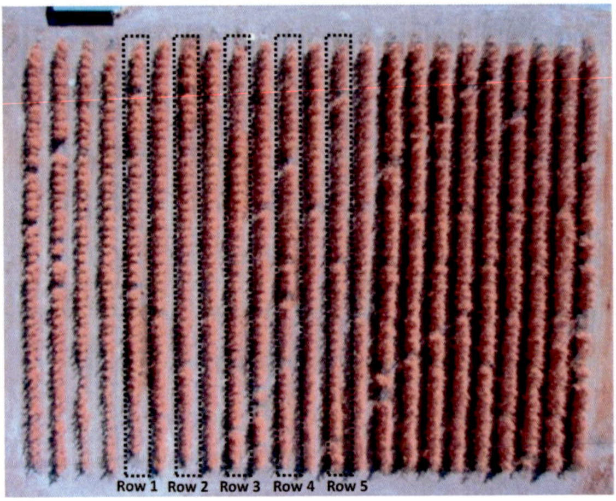

Figure 4.4. *Sample multispectral image of the target orchard.*

Source: Bulanon, D. M., Lonai, J., Skovgard, H., & Fallahi, E. (2016). Evaluation of different irrigation methods for an apple orchard using an aerial imaging system. ISPRS International Journal of Geo-Information, 5(6), 79. https://doi.org/10.3390/ijgi5060079.

4.5.1. Applications in Insect Pest Management

In recent times, remote sensing (RS) has become increasingly important in pest management. RS data can be obtained either through airborne methods with lower spatial resolution for vast area coverage, or through high spatial resolution techniques. RS has emerged as a crucial tool in studying insect physiology, behaviors, and ecology. Wang et al. (2016) demonstrated the practicality of using hyperspectral imaging data for insect classification. Integrative taxonomy, a modern approach to insect species classification, helps identify cryptic and closely related species by combining morphological, ecological, and genetic data.

Through a comprehensive analysis of mitochondrial DNA, morphological characteristics, and hyperspectral reflectance, seven species belonging to the evacanthine leafhopper genus *Bundera* Distant (Hemiptera: Cicadellidae) were identified. The hyperspectral reflectance profiling used 37 spectral bands ranging from 411 to 870 nm. The reflectance-based classification data achieved an impressive 91.3% accuracy in distinguishing the seven species (Shimoda & Honda, 2013).

In addition, these techniques foster interdisciplinary collaboration, facilitating interaction between entomologists and experts in fields such as electrical and informatics engineering. RS techniques can reveal insect activity by assessing the damage due to the target insect pest, rather than directly detecting the organism itself. Damage levels are typically identifiable through the presence of sooty mold deposits caused by insects that produce honeydew, the defoliation of plants, shifts in color, and finally, the geometric distortion in the plant shape. RS sensors have become valuable tools for monitoring migration swarms, pest scouting, outbreak prediction, detecting infestation levels, and pest management which invade various crops and fruit orchards (Shelton & Badenes, 2006). These sensors have been extensively studied and utilized by experts in the field.

In addition, spatial statistics tools are employed in RS to discover and model the various spatial correlations and links between the environmental and climatic factors. For example, a study conducted in The Sultanate of Oman mapped and modeled the population density and habitat of the *Ommatissus lybicus* Bergevin (Hemiptera: Tropiduchidae) using RS and spatial statistics. Various methods were employed to monitor insect pests, including the use of optoelectronic and optical devices like videography, thermal imaging, radiotelemetry, radio fre□uency identification (RFID), radiography, X-rays, radar, satellites, and computed tomography. There have been numerous successful applications reported in insect infestation detection using airborne RS. For instance, brown soft scale and citrus blackfly issues, along with whitefly infestation in cotton, were identified through the use of multi-spectral videography and aerial CIR (color-infrared) photography to perceive infestations on trees throughout certain citrus orchards (Everitt, 1996).

The identification of early mite infestation in large-scale cotton fields was further investigated by the use of coarse-grained spatial resolution multi-spectral RS (MRS). It primarily depends on color and time alterations in the canopy (Fitzgerald, 2000). In addition, an on-ground hyperspectral RS system was utilized in pest management, as demonstrated in studies conducted by Prabhakar et al. (2011). For instance, Prabhakar et al. (2011) detected damage caused by leafhoppers in cotton fields. A hyperspectral radiometer, utilizing a 350–2500 nm spectral range is employed to determine the relative water content and chlorophyll content of certain plants. There are noticeable variations in reflectance measurements between healthy plants and those infested with leafhoppers, particularly in the NIR and VIS regions. The infested plants exhibit a noticeable decrease in chlorophyll A pigment compared to chlorophyll B pigment. In addition, NIR spectroscopy shows great potential as a method for identifying insects that infest stored food products. It also allows for quick, cost-effective, and non-harmful analysis of the cuticular hydrocarbons found on individual insects (Johnson & Naiker, 2019).

RS technology has been utilized in Egypt as a valuable tool to detect and monitor the distribution of *N. bruchi* Hustache and *Neochetina eichhorinae* Warner weevils, which are biological control agents for water hyacinth. This has been achieved through the use of multi-spectral SPOT4 or Satellite Probatoire de l'Observation de la Terre, RS imaging technique in Mariout and Edko lakes (Fayad et al., 2003). Using this imaging, an RS technique was utilized in the lake to detect water hyacinth infestation both pre-and post-release of *N. bruchi* Hustache and *N. eichhorinae* Warner weevils. After just one year, a noticeable decrease in the infestation of water hyacinths was observed. The water hyacinth in Mariout and Edko and Mariout lakes has decreased by 34.35% and 30%, respectively. In a later study, Hassan et al. (2003) implemented GIS maps to track insect vectors of filariasis and malaria.

Entomological radar's remote sensing (RS) capabilities allow for the tracking and monitoring of pests harmful to agriculture, medicine, and forestry, as well as their distribution. It is used as an alternative to traditional insect sampling methods like sweep nets and traps. Radar offers significant advantages, such as its ability to detect insects at altitudes exceeding 1 km above ground level (AGL) and its capacity to sample a larger volume compared to conventional methods (Chapman et al., 2002). Additionally, radar provides a unique opportunity to study various aspects of insect migration behavior, including the orientation of high-flying migrants. Radar is commonly employed in insect migration research due to its ability to monitor nocturnal insects.

Different types of entomological radars include tracking entomological radar (TER), scanning entomological radar (SER), vertical-looking radar (VLR), and airborne entomological radar (AER). Among these, VLR and SER are widely used to study the migration behavior of high-altitude insects. A typical entomological radar operates on a 9.4 GHz X-band frequency and uses a vertical pencil beam. VLR provides precise measurements of aerial insect density, biomass, and diversity over different time scales, making it a valuable tool for tracking changes in insect population dynamics.

However, the limited recurrence rate of current insect radars has hindered the acquisition of high-resolution data needed for identifying individual insects. Hu et al. (2019) conducted a study in East China using multi-frequency radar to investigate 15 insect pests from 3 families and 11 species, examining their recurrence patterns. Using multi-frequency radar cross-sections, they were able to estimate insect body length and mass with uncertainties of 10.74% and 16.31%, respectively. Estimates for the thorax aspect and width ratio had uncertainties of 7.99% and 13.37%, respectively. After analyzing statistical data from 5,532 insects across 23 different species, it was determined that the likelihood of correctly identifying all 23 species was greater than 0.5. For 15 of these species, the identification odds exceeded 0.8, demonstrating the high accuracy of the technique (Rai & Ingle, 2012).

Populations of migrating insects can be tracked using entomological radar. Observing the movements of migratory insects is particularly challenging due to their small body size, nocturnal activity, and high-altitude flights. Studying insect migration presents significant limitations when relying on traditional methods such as light traps and other ground-based observations, especially for tracking high-altitude movement (Berger, 1977). Additionally, maintaining surveillance systems, such as drones or tethered balloons, can be expensive and impractical over extended periods.

In a groundbreaking study, a harmonic radar system was introduced, capable of tracking high-altitude flying insects, specifically *Vespa velutina* Lepeletier, over a range of 500 meters. This advanced technology provides a wide field of view in terms of elevation, making it a valuable tool for tracking insect flight paths and locating nests that need to be eradicated to control *V. velutina* populations (Khater, 2012).

Numerous RS technologies, such as entomological radars and the NDVI satellite indicator, were used to monitor and forecast numerous species of locusts (Smith et al., 1993). This insect can cause significant damage to cultivated areas, particularly during migration and swarming over long distances. Just one swarm may cover an impressive area of 1200 km^2 and is home to a staggering number of locusts, ranging from 40 to 80 million per km^2. Over 8 million individuals were impacted by the plague that occurred in West Africa between 2003 and 2005. Significant losses were observed in pasture (85%); legumes (90%); and cereals (100%). Managing the plague required nearly $600 million and almost 13 million liters of pesticides. This was necessary due to the speedy vegetation growth that followed a cycle of drought, leading to locust swarms. Various satellites can track factors like soil moisture and greenery that might lead to locust swarms and aid in the prediction of locust plagues. Entomological radar data records the body shape and size, along with the measurement of wing-beat frequency. These measurements are crucial in enabling the radar to effectively distinguish desert locusts from various high-flying insects. Using satellite information, researchers have found that it is a promising tool for managing species like the desert and Australian locusts (Sivanpillai & Latchininsky, 2013).

Recent collaboration between ESA's satellite and international collaborators from the United Nations Food and Agriculture Organization (FAO), Spain, Mali, Mauritania, Morocco, France, and Algeria aims to forecast locust plagues. In addition, RS has proven to be an effective tool for monitoring, mapping, and controlling damage to forests from insects. The said aerial sketch mapping techni□ue is highly efficient in assessing forest health through the use of color-infrared and color photos taken from low-flying aircraft. In addition, various RS tools were utilized, including earth-observing satellites, digital camera systems, and airborne video. Using these RS technologies is a good way to keep an eye on and lessen the growing problems caused in various conifer forests by the following beetles: Douglas fir (*D. pseudotsugae* Hopkins), spruce (*D. rufipennis* Kirby), mountain pine (*Dendroctonus ponderosae* Hopkins), southern pine (*D. frontalis* Zimmermann), bark beetle (*Scolytus* sp.), and others. Meigs et al. (2011) provide a clear example of the correlation between pest damages and field-based measures, such as bark beetle infestations, and changes in Landsat spectral data. They applied the LandTrend algorithm across the Cascade Range in Oregon, USA, to assess the impact of bark beetle infestation on tree mortality. Their analysis revealed a strong correlation between field-based measurements and spectral data from Landsat (Khater, 2012).

Various RS technologies, including satellite imaging systems like Landsat, SPOT, MODIS, and AVHRR, have been used to estimate population fluctuations and map disease vector insects such as mosquitoes (*Aedes* sp.; *Culex* sp.; and *Anopheles* sp.,) and tsetse flies (*Glossina morsitans* Wiedemann). In addition, these skills were used to address insect pests in the field crops. For instance, RS technology has proven effective in detecting the presence of insect pests such as the *Rhopalosiphum maidis* Fitch the corn leaf aphid, and the *Bemicia tabaci* Glover, or the sweet potato whitefly. This is done by identifying the visible sooty mold that grows on the honeydew these pests produce. In a case study, high-quality images captured from varying altitudes (300 to 2000 m above) revealed clear patterns of mold growth in cotton fields (Acharya & Thapa, 2015).

Social insect pests have been added to the list of entomology-related RS applications. These tools have been used to study different species of termites, including both non- and subterranean species. These insect pests are economically significant due to their extensive damage to historical structures across the globe, homes, and timber plantations. Termite infestation can be detected using millimeter wave imaging and acoustic signal techniques. The said techni□ue relies on the reflected energy from the body when exposed to the non-ionizing EM millimeter waves. The waves are emitted at a 100 GHz frequency with a 10-mW output power (Yoshihisa et al., 2007) (Figure 4.5).

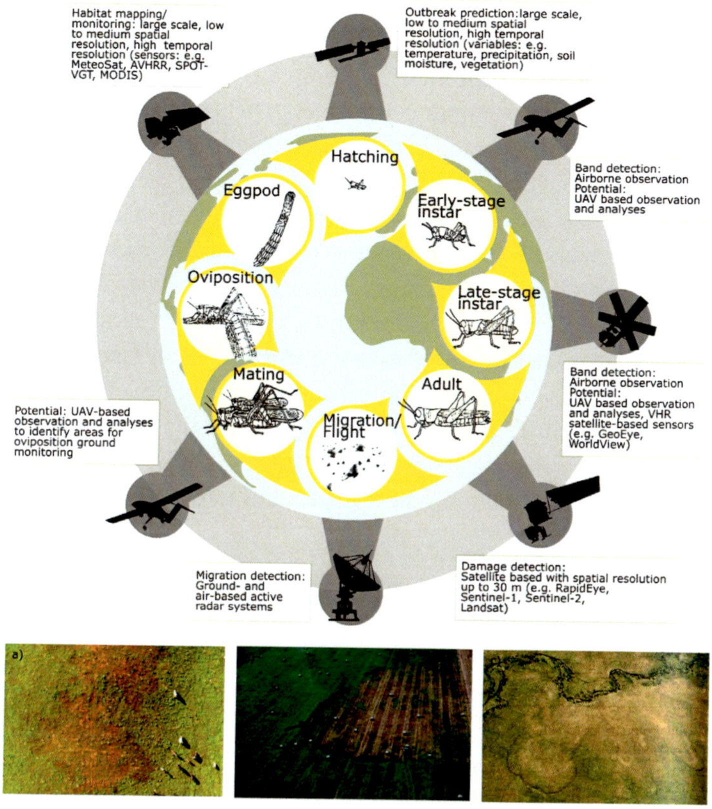

Figure 4.5. *Locating the locust infestation.*

Source: Klein, I., Oppelt, N., & Kuenzer, C. (2021). Application of remote sensing data for locust research and management – A review. Insects, 12(3), 233. https://doi.org/10.3390/insects12030233.

4.5.2. Applications in Plant Disease Management

The protection of crops from plant infections is of utmost importance. Modern issues are met with the logical investigation into agricultural handicrafts. Practical and environmentally friendly solutions are increasingly in demand (González et al., 2023).

Accurately identifying the initial sites of infection and understanding how diseases spread is crucial for determining the best course of action for managing them. Optical sensors provide an accurate and unbiased detection of plant infections in this situation. Plant phenotyping heavily relies on optical sensor-based approaches, which are widely recognized as crucial (Khan & Rizvi, 2014).

When the sun's EM energy interacts with plants, there are three possible outcomes: 1) reflection; 2) absorption; or 3) transmission. RS can identify the energy that is reflected, transmitted, or absorbed based on the properties of specific plants and the wavelength of the energy. The amount of energy that is transmitted, absorbed, or reflected can explain variations in leaf colors, textures, forms, and even how the leaves are linked to the plants. Higher NIR often denotes healthy plants, at least one of which is visible; infected plants, on the other hand, exhibit the reverse behavior. Plant stress typically leads to an elevation in the visible reflectance of its surface. This occurs as the chlorophyll level decreases, resulting in reduced absorption of visible light (Rani et al., 2018). In 1933, Bawden conducted a study on the use of visible aerial photography to detect viral diseases in potato and tobacco crops. Colwell (1956) utilized infrared imaging to identify disease-related changes in the internal leaf structure of cereal crops. Following that, these methods were applied to a wide range of crop diseases, including sheath blight on rice, cotton root rot, bacterial blight of beans, and potato blight (Kumar et al., 2022).

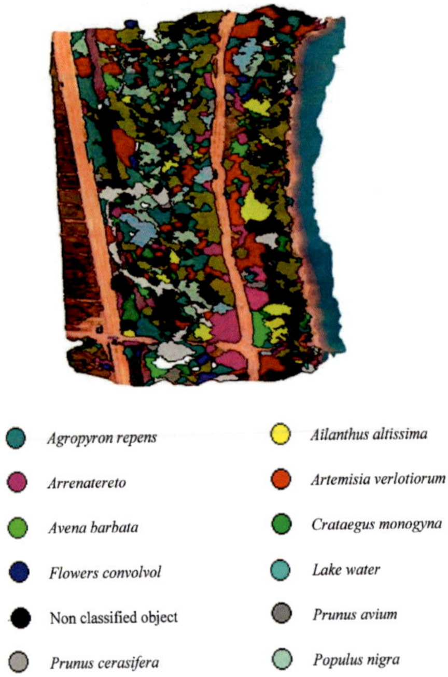

Figure 4.6. *A general spectral curve is shown by a healthy plant.*

Source: Tariku, G., Ghiglieno, I., Gilioli, G., Gentilin, F., Armiraglio, S., & Serina, I. (2023). Automated identification and classification of plant species in heterogeneous plant areas using unmanned aerial vehicle-collected RGB images and transfer learning. Drones, 7(10), 599. https://doi.org/10.3390/drones7100599.

In recent years, spectral sensors have been increasingly used to enhance site-specific disease management. For instance, hyperspectral technology was used to assess the detection and quantification of yellow rust disease in wheat crops (Kuska & Mahlein, 2018). Field experiments involve the use of a hyperspectral camera mounted on two different measurement platforms: a ground-based vehicle and a UAV. These platforms enable the differentiation of various stages and intensities of yellow rust disease in the wheat fields. The light reflected by pathogens and diseases that develop on crop leaves is measured by spectral sensors, which create a spectral fingerprint in the reflected leaf signature. These changes in the crop canopy's signature can be identified by using spectral sensors, specifically in the EM spectrum ranging from 400–2500 nm (Nelson et al., 1999) (Figure 4.6).

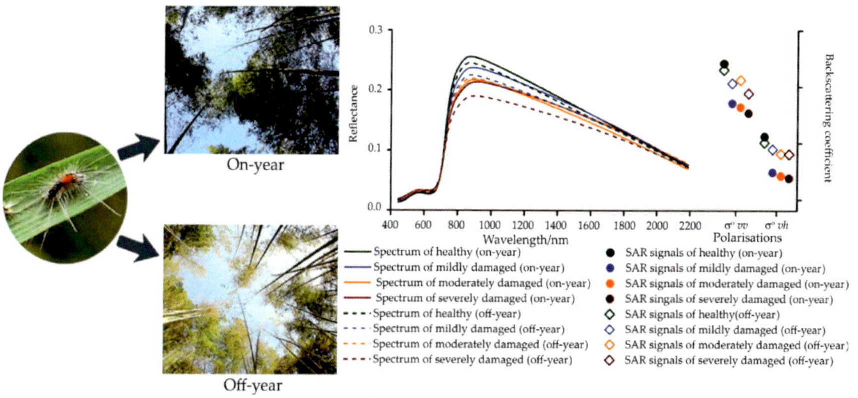

Figure 4.7. *Spectral profiles of Pantana phyllostachysae Chao (PPC) infested bamboo leaves.*

Source: Huang, X., Zhang, Q., Hu, L., Zhu, T., Zhou, X., Zhang, Y., Xu, Z., & Ju, W. (2022). Monitoring damage caused by Pantana phyllostachysae Chao to Moso bamboo forests using Sentinel-1 and Sentinel-2 images. Remote Sensing, 14(19), 5012. https://doi.org/10.3390/rs14195012.

In addition, aerial imaging and satellite devices with high spatial resolution have been utilized to study crop disease issues and evaluate their effects on productivity. Several studies have focused on this area. When it comes to applying crop protection and production materials, the resulting image can offer valuable insights that can be used to generate prescription data. RS utilizes a variety of tools, including satellites, aircraft, and ground-based platforms (Van Bruggen et al., 2016). For example, the Landsat 5 TM image is used to find and measure the amount of site-specific wheat streak mosaic disease in wheat crops. In addition, aerial photography was utilized to create forest cover maps and quantify the number of trees that were harvested. In addition, it has been utilized in forest trees to identify diseases. Area-wide pest management (APM), also known as reactive field-by-field control, is essentially the idea of preventing the spread of a particular species of pest throughout its geographic range. Airborne RS has been widely utilized for detecting and analyzing the distribution of weeds (Huang et al., 2008). Radiance ratio and NDVI techniques have been employed recently to discriminate between solid weed and pure wheat plots of dwarf mallow, *Malva neglecta* Wallr (Martin et al., 2000) (Figure 4.7).

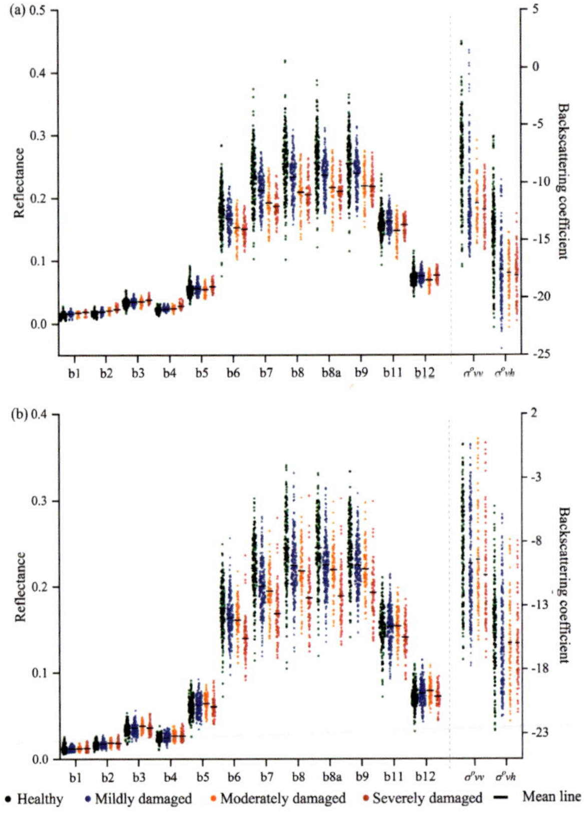

Figure 4.8. *The spectra and SAR signals of (a) on-year; and (□) off-year □am□oo forests with different damage severities. b1~b9 represent the coastal, blue, green, red, red-edge$_{705}$, red-edge$_{740}$, red-edge$_{783}$, NIR, NIR$_{narrow}$, and water vapor bands of sentinel-2 imagery, while b11~b12 represent the SWIR-1 and SWIR-2 bands of sentinel-2 imagery.*

Source: Huang, X., Zhang, Q., Hu, L., Zhu, T., Zhou, X., Zhang, Y., Xu, Z., & Ju, W. (2022). Monitoring damage caused by Pantana phyllostachysae Chao to Moso bamboo forests using Sentinel-1 and Sentinel-2 images. Remote Sensing, 14(19), 5012. https://doi.org/10.3390/rs14195012.

Infected maize was evaluated on the Ofcolaco farm in the Limpopo Province of Tzaneen, South Africa, to identify the Maize Streak Gemini virus (MSV) using additional detection techniques. Based on the guided regularized random forest (GRRF) algorithm, the ability of hyperspectral information to classify distinct levels of MSV contamination in maize was examined. The GRRF algorithm is highly effective in identifying and selecting the optimal wavelengths needed to differentiate between different levels of MSV crop diseases. Based on the GRRF algorithm, the most effective groups for identifying different levels of maize streak infection were found to be at wavelengths of 552 nm, 603 nm, 683 nm, 881 nm, and 2338 nm. This study emphasizes the importance of using remotely sensed data to accurately detect and assess crop infections such as MSV. This is crucial for ensuring food safety, particularly in regions with limited resources like sub-Saharan Africa (Coakley et al., 1999). Utilizing the RapidEye multi-spectral sensor and vegetation indices, we were able to effectively recognize and map the presence of MSV in infected maize (Figure 4.8). It turns out that the utilization of RapidEye satellite imagery in the mapping and detection of MSV resulted in a significant accuracy rate of 82.75%. Ultimately, the utilization of satellite-based VIs significantly improved the accuracy of classification. The RapidEye horrifying group classification handle showed 82.75%, and the computed vegetation file records improved the classification accuracy by 3.4% (Dhau et al., 2018).

4.6. SUMMARY

Ultimately, the recent progress in RS technology has opened up new possibilities for various agricultural practices, particularly in the areas of pest control and plant disease management. With the advancements in data analysis, interpretation, and information technology, high-tech applications have brought about numerous benefits in various research fields, aiding in the growth and support of knowledge. RS has become increasingly valuable as a versatile tool for a wide range of users with varying needs. RS is a highly promising technology that supports integrated pest management by allowing the investigation of various environmental parameters and natural sources as a comprehensive system. However, there is a growing need for the development of RS systems that are more cost-effective and produce higher-quality imaging.

REFERENCES

1. Acharya, M. C., & Thapa, R. B. (2015). Remote sensing and its application in agricultural pest management. *Vol. 1,* (pp. 41–61).
2. Andreo, P., Wulff, J., Burns, D. T., & Palmans, H. (2013). Consistency in reference radiotherapy dosimetry: Resolution of an apparent conundrum when 60Co is the reference quality for charged-particle and photon beams. *Physics in Medicine & Biology, 58*(19), 6593.
3. Banerjee, T., Singh, M. K., & Khanna, A. (2006). Prediction of binary VLE for imidazolium-based ionic liquid systems using COSMO-RS. *Industrial & Engineering Chemistry Research, 45*(9), 3207–3219.
4. Berger, R. D. (1977). Application of epidemiological principles to achieve plant disease control. *Annual Review of Phytopathology, 15*(1), 165–181.
5. Chapman, M. R., Robinson, L. S., Pinkner, J. S., Roth, R., Heuser, J., Hammar, M., & Hultgren, S. J. (2002). Role of *Escherichia coli* curli operons in directing amyloid fiber formation. *Science, 295*(5556), 851–855.
6. Cheng, C. H., & Chen, Y. S. (2009). Classifying the segmentation of customer value via the RFM model and RS theory. *Expert Systems with Applications, 36*(3), 4176–4184.
7. Coakley, S. M., Scherm, H., & Chakraborty, S. (1999). Climate change and plant disease management. *Annual Review of Phytopathology, 37*(1), 399–426.
8. Colwell, L. V. (1956). The effects of high-frequency vibrations in grinding. *Transactions of the American Society of Mechanical Engineers, 78*(4), 837–844.
9. Dhau, I., Adam, E., Mutanga, O., & Ayisi, K. K. (2018). Detecting the severity of maize streak virus infestations in maize crops using *in situ* hyperspectral data. *Transactions of the Royal Society of South Africa, 73*(1), 8–15.
10. Dong, A. Y., Wang, Z., Huang, J. J., Song, B. A., & Hao, G. F. (2021). Bioinformatic tools support decision-making in plant disease management. *Trends in Plant Science, 26*(9), 953–967.
11. Eigenbrode, S. D., Birch, A. N. E., Lindzey, S., Meadow, R., & Snyder, W. E. (2016). A mechanistic framework to improve understanding and applications of push-pull systems in pest management. *Journal of Applied Ecology, 53*(1), 202–212.
12. Everitt, A. (1996). Developing critical evaluation. *Evaluation, 2*(2), 173–188.

13. Fayad, J. N., Wanna, G. B., Micheletto, J. N., & Parisier, S. C. (2003). Facial nerve paralysis following cochlear implant surgery. *The Laryngoscope, 113*(8), 1344–1346.
14. Fernando, T., Denman, S., Sridharan, S., & Fookes, C. (2018). Soft+ hardwired attention: An LSTM framework for human trajectory prediction and abnormal event detection. *Neural Networks, 108,* 466–478.
15. Fitzgerald, R. (2000). Phase-sensitive X-ray imaging. *Physics Today, 53*(7), 23–26.
16. Freire, M. G., Ventura, S. P., Santos, L. M., Marrucho, I. M., & Coutinho, J. A. (2008). Evaluation of COSMO-RS for the prediction of LLE and VLE of water and ionic liquids binary systems. *Fluid Phase Equilibria, 268*(1–2), 74–84.
17. González-Domínguez, E., Caffi, T., Rossi, V., Salotti, I., & Fedele, G. (2023). Plant disease models and forecasting: Changes in principles and applications over the last 50 years. *Phytopathology®, 113*(4), 678–693.
18. Hall, R. J., Castilla, G., White, J. C., Cooke, B. J., & Skakun, R. S. (2016). Remote sensing of forest pest damage: A review and lessons learned from a Canadian perspective. *The Canadian Entomologist, 148*(S1), S296–S356.
19. Hassan, M. K., & Bashir, A. H. M. (2003, December). Determinants of Islamic banking profitability. In *10th ERF Annual Conference, Morocco,* Vol. 7, (pp. 2–31).
20. Hu, C., Zhang, L., & Gong, J. (2019). Recent progress made in the mechanism comprehension and design of electrocatalysts for alkaline water splitting. *Energy & Environmental Science, 12*(9), 2620–2645.
21. Huang, J., Liu, Y., Hou, H., & You, T. (2008). Simultaneous electrochemical determination of dopamine, uric acid, and ascorbic acid using palladium nanoparticle-loaded carbon nanofibers modified electrode. *Biosensors and Bioelectronics, 24*(4), 632–637.
22. Johnson, J., Collins, T., Skylas, D., & Naiker, M. (2019, August). ATR-MIR: A valuable tool for the rapid assessment of biochemically active compounds in grains. In *69th Australasian Grain Science Conference, Carlton, Melbourne, Australia,* 2(1), (pp. 27–29).
23. Khan, M. R., & Rizvi, T. F. (2014). Nanotechnology: Scope and application in plant disease management. *Plant Pathology Journal, 13*(3), 214–231.
24. Khater, H. F. (2012). Prospects of botanical biopesticides in insect pest management. *Pharmacologia, 3*(12), 641–656.

25. Kumar, A., Choudhary, A., Kaur, H., Guha, S., Mehta, S., & Husen, A. (2022). Potential applications of engineered nanoparticles in plant disease management: A critical update. *Chemosphere, 295,* 133798.

26. Kuska, M. T., Behmann, J., & Mahlein, A. K. (2018). The potential of hyperspectral imaging to detect and identify the impact of chemical warfare compounds on plant tissue. *Pure and Applied Chemistry, 90*(10), 1615–1624.

27. Lanza, A. F. (2005). On the orbital period modulation of RS CVn binary systems. *Monthly Notices of the Royal Astronomical Society, 364*(1), 238–246.

28. Malinowski, D. (2015). Opening spaces of learning in the linguistic landscape. *Linguistic Landscape, 1*(1–2), 95–113.

29. Malinowski, P., & Lim, H. J. (2015). Mindfulness at work: Positive affect, hope, and optimism mediate the relationship between dispositional mindfulness, work engagement, and well-being. *Mindfulness, 6,* 1250–1262.

30. Martin, R. R., James, D., & Lévesque, C. A. (2000). Impacts of molecular diagnostic technologies on plant disease management. *Annual Review of Phytopathology, 38*(1), 207–239.

31. Meigs, G. W., Kennedy, R. E., & Cohen, W. B. (2011). A Landsat time series approach to characterize bark beetle and defoliator impacts on tree mortality and surface fuels in conifer forests. *Remote Sensing of Environment, 115*(12), 3707–3718.

32. Moo, P. W., & DiFilippo, D. J. (2018). Multifunction RF systems for naval platforms. *Sensors, 18*(7), 2076.

33. Nelson, M. R., Orum, T. V., Jaime-Garcia, R., & Nadeem, A. (1999). Applications of geographic information systems and geostatistics in plant disease epidemiology and management. *Plant Disease, 83*(4), 308–319.

34. Nuzzo, P., Sangiovanni-Vincentelli, A. L., Bresolin, D., Geretti, L., & Villa, T. (2015). A platform-based design methodology with contracts and related tools for the design of cyber-physical systems. *Proceedings of the IEEE, 103*(11), 2104–2132.

35. Oumar, Z., & Mutanga, O. (2011). The potential of remote sensing technology for the detection and mapping of *Thaumastocoris peregrinus* in plantation forests. *Southern Forests: A Journal of Forest Science, 73*(1), 23–31.

36. Palomar, J., Gonzalez-Miquel, M., Polo, A., & Rodriguez, F. (2011). Understanding the physical absorption of CO_2 in ionic liquids using the

COSMO-RS method. *Industrial & Engineering Chemistry Research, 50*(6), 3452–3463.

37. Peralta, G., & Zareei, A. (2016). A network approach to portfolio selection. *Journal of Empirical Finance, 38,* 157–180.
38. Prabhakar, P. K., & Doble, M. (2011). Mechanism of action of natural products used in the treatment of diabetes mellitus. *Chinese Journal of Integrative Medicine, 17,* 563–574.
39. Prasad, Y. G., Prabhakar, M., Sreedevi, G., Rao, G. R., & Venkateswarlu, B. (2012). Effect of temperature on development, survival, and reproduction of the mealybug, *Phenacoccus solenopsis* Tinsley (Hemiptera: Pseudococcidae) on cotton. *Crop Protection, 39,* 81–88.
40. Rai, M., & Ingle, A. (2012). Role of nanotechnology in agriculture with special reference to management of insect pests. *Applied Microbiology and Biotechnology, 94,* 287–293.
41. Rani, M., Dhok, S. B., & Deshmukh, R. B. (2018). A systematic review of compressive sensing: Concepts, implementations, and applications. *IEEE Access, 6,* 4875–4894.
42. Ricci, F. (2010). Mobile recommender systems. *Information Technology & Tourism, 12*(3), 205–231.
43. Rogers, D. J., & Randolph, S. E. (2003). Studying the global distribution of infectious diseases using GIS and RS. *Nature Reviews Microbiology, 1*(3), 231–237.
44. Shelton, A. M., & Badenes-Perez, F. R. (2006). Concepts and applications of trap cropping in pest management. *Annual Review of Entomology, 51,* 285 308.
45. Shimoda, M., & Honda, K. I. (2013). Insect reactions to light and its applications to pest management. *Applied Entomology and Zoology, 48,* 413–421.
46. Sivanpillai, R., & Latchininsky, A. V. (2013). Publisher's note: Special section on advances in remote sensing applications for locust habitat monitoring and management. *Journal of Applied Remote Sensing, 7*(1), 075001.
47. Smith, C. A., Haynes, K. N., Lazarus, R. S., & Pope, L. K. (1993). In search of the "hot" cognitions: Attributions, appraisals, and their relation to emotion. *Journal of Personality and Social Psychology, 65*(5), 916.
48. Van Bruggen, A. H., Gamliel, A., & Finckh, M. R. (2016). Plant disease management in organic farming systems. *Pest Management Science, 72*(1), 30–44.

49. Wang, J., & Wang, S. (2016). Removal of pharmaceuticals and personal care products (PPCPs) from wastewater: A review. *Journal of Environmental Management, 182,* 620–640.

50. Watts, A. C., Ambrosia, V. G., & Hinkley, E. A. (2012). Unmanned aircraft systems in remote sensing and scientific research: Classification and considerations of use. *Remote Sensing, 4*(6), 1671–1692.

51. White, N. E., Sanford, P. W., & Weiler, E. J. (1978). An X-ray outburst from the RS CVn binary HR1099. *Nature, 274*(5671), 569.

52. Yan, M., Kawamata, Y., & Baran, P. S. (2017). Synthetic organic electrochemical methods since 2000: On the verge of a renaissance. *Chemical Reviews, 117*(21), 13230–13319.

53. Yang, C., Everitt, J. H., & Murden, D. (2011). Evaluating high-resolution SPOT 5 satellite imagery for crop identification. *Computers and Electronics in Agriculture, 75*(2), 347–354.

54. Yang, J. J., Miao, F., Pickett, M. D., Ohlberg, D. A., Stewart, D. R., Lau, C. N., & Williams, R. S. (2009). The mechanism of electroforming of metal oxide memristive switches. *Nanotechnology, 20*(21), 215201.

55. Yoshihisa, T., Ohshima, C., Yunoki-Esaki, K., & Endo, T. (2007). Cytoplasmic splicing of tRNA in *Saccharomyces cerevisiae. Genes to Cells, 12*(3), 285–297.

56. Zhan, J., Thrall, P. H., & Burdon, J. J. (2014). Achieving sustainable plant disease management through evolutionary principles. *Trends in Plant Science, 19*(9), 570–575.

57. Zhang, C., & Kovacs, J. M. (2012). The application of small unmanned aerial systems for precision agriculture: A review. *Precision Agriculture, 13,* 693–712.

58. Zhang, J., Huang, Y., Pu, R., Gonzalez-Moreno, P., Yuan, L., Wu, K., & Huang, W. (2019). Monitoring plant diseases and pests through remote sensing technology: A review. *Computers and Electronics in Agriculture, 165,* 104943.

59. Zrelli, A. (2022). Hardware, software platforms, operating systems, and routing protocols for Internet of Things applications. *Wireless Personal Communications, 122*(4), 3889–3912.

Chapter **5**

ROLE OF IMAGES IN REMOTE SENSING

Contents

5.1. INTRODUCTION

There are two distinct perspectives when it comes to satellite imagery: one sees it as a captured photograph, while the other views it as a visual representation of data on spectral intensity that measures the reflection of light from objects on a plant's surface. When utilizing satellite imagery, the focus is often on enhancing and emphasizing the information within the image. This can be achieved through various tools and techniques, similar to how a skilled photographer would enhance their photos in a darkroom or using digital editing software like Photoshop. Using automated processing techniques found in remote sensing software that depends on a system of equations to quantify data about reflected light, another goal might be to alter the image. The objective of both methods is to gather data about the observed ground conditions (Li et al., 2014).

At first look, the picture in Figure 5.1 does not look much like a landscape that a majority would identify as being on Earth. Indeed, the colors that stand out the most are bright turquoise and orange. In addition to allowing for a wide range of objects to be recognized on the ground, the use of color in visual imagery elevates picture interpretation to the level of art. Even someone new to interpreting can make sense of the image, while those with more knowledge and experience of its color scheme can discern even finer details. In Figure 5.1, there are several notable features to observe. On the left, there is a river indicated by a blue line. Throughout the image, there is a range of vegetation displayed in various shades of orange, transitioning from light to dark. Additionally, there are turquoise patches that represent burn scars (Bauer, 1975).

Figure 5.2 depicts a section of the landscape that is seen in Figure 5.1's satellite image. The location of the photograph is indicated by the red dot in Figure 5.1. From the vantage point indicated by the red dot, this photograph captures the view towards the south, as it would appear to a human observer. There are two significant differences between the photograph viewpoints and the satellite image. One important factor to consider is the viewpoint where the images were taken. While the satellite image was obtained using a sensor in the orbit staring vertically down at Earth, the photograph was taken with the camera positioned slightly horizontally (Klemas, 2009). It is evident that the photograph captures significantly more landscape detail compared to the satellite image. The satellite image offers a more abstract perspective of the landscape in contrast to the photograph. The capacity to correlate what can be seen in an image with actual physical elements on the ground is perhaps the most crucial skill when using remotely sensed imagery. This chapter will delve into the details of working with remotely sensed imagery, enabling the reader to develop their skills in interpreting imagery and recognizing various landscape features (Brisco et al., 1998).

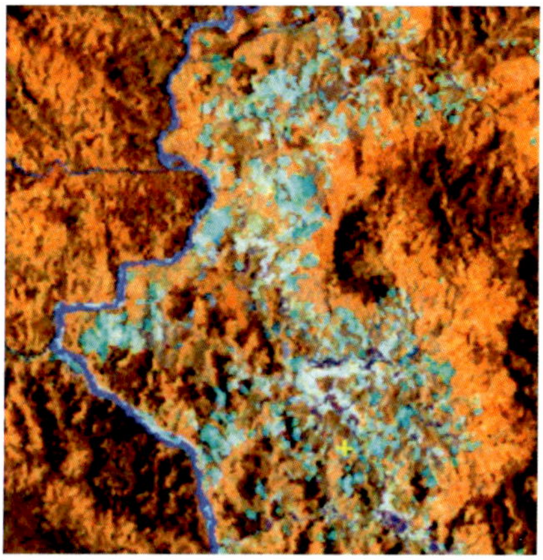

Figure 5.1. *An area in Quang Nam Province, central Vietnam, is covered by the Landsat ETM+ image.*

Source: MDPI; Creative Commons License.

Figure 5.2. *A view of a landscape taken in April 2002 in the central Vietnamese province of Quang Nam.*

Source: MDPI; Creative Commons License.

5.2. CHOOSING THE RIGHT TYPE OF IMAGERY

There are two types of remote sensors: active and passive. One key distinction between the two categories is the energy source utilized to illuminate ground features. Active sensors generate their energy, while passive sensors depend on an external source of energy, typically the sun. There are two commonly used active sensors in the field of remote sensing. One is radar, which stands for radio detection and ranging. It works by measuring the return of radio waves to a sensor. The other is lidar, which stands for light detection and ranging. It gauges the return of light from a laser. While the frequency of data obtained from active sensors is rising, most projects still rely on imagery obtained from passive sensors. This is mainly due to the abundance of passive RS imagery, which is easily accessible and can be processed and visualized with ease (Rosenbaum, 2021).

The Operational Linescan System or OLS by the United States Air Force Defense Meteorological Satellite Program (DMSP), which has been used since 1994 to create a global nighttime lights product, is one example of an operational passive satellite-based sensor that gathers data during the night. It is highly responsive to faint lights, particularly those generated in urban environments. The National Geophysical Data Center NGDC NOAA distributes nighttime light products, which are useful for estimating the density and increase of the worldwide human population (Chang et al., 2018).

Passive RS devices commonly used are cameras, which do not have a flash. When it comes to digital cameras, an electronic image plate, that is, a sensor is used as the medium of recording, whereas film cameras use film as the medium for capturing images. Regardless of the scenario, the camera captures the light's intensity that is reflected from a particular object, like a landscape, producing an image that closely resembles how the human eye perceives it (Hedblom et al., 2015).

5.2.1. Multispectral Imaging

Remote sensing instruments, such as cameras, can collect data in various bands or layers. Different bands capture a specific range of wavelengths. With a common color camera, data is recorded in three layers, red, blue, and green or RGB, representing the primary colors of light. Because these light wavelengths correspond to colors that the human eye can see, a color photograph closely resembles the element that was photographed (Levenson & Mansfield, 2006).

More advanced remote sensing instruments, like those found in satellites, can acquire multiple bands. These bands can detect UV and infrared wavelengths of light that are outside the range of human vision. Hyperspectral imagers capture data across numerous bands, ranging from dozens to even hundreds. These sensors enhance the capacity to identify various features on the Earth's surface by providing multiple bands. Being able to observe wavelengths of light beyond what the human eye can see gives extra information to more accurately analyze important features (Dinguirard & Slater, 1999).

Figure 5.3 displays the previous Landsat image with various band combinations. The initial image was generated using bands that mimic those of a digital camera. This means that the band with the closest wavelength to red is shown as red, the band with the wavelength closest to green is shown as green, and the band with the closest wavelength to blue is shown as blue. The band combinations are usually presented in the sequence of red, green, and blue, with a distinct number allocated to every band, specific to the remote sensing instrument being used.

Additional images were generated by utilizing various band combinations with data from the NIR and visible spectrum range (Shogenji et al., 2004). It is possible to display any band using the three primary colors, although only a maximum of three bands will be displayed simultaneously. Take note of how some band combinations work better to highlight distinct aspects in Figure 5.3. There is no foolproof method for determining the perfect band combination. Experimenting with different band sequences is often the most effective approach to finding the combination that suits the specific needs and preferences for the required features. After gaining expertise in any equipment along its bands, interpreters often discover their preferred display combinations for specific purposes (Pelagotti et al., 2008).

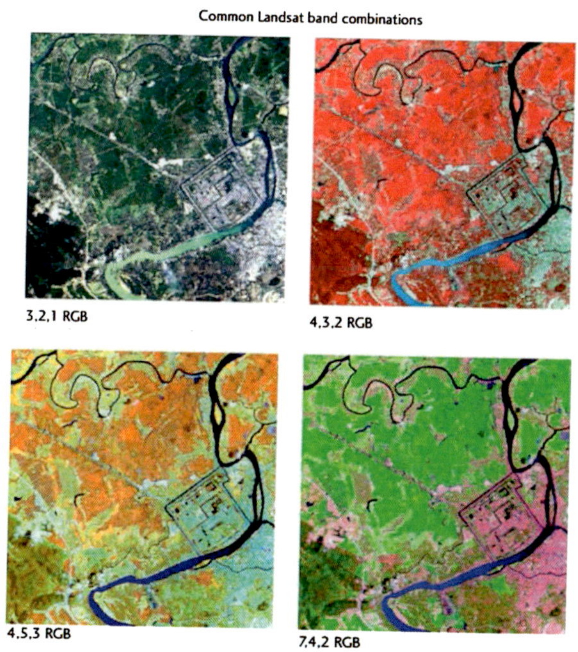

Figure 5.3. *Typical band configurations for Landsat ETM+ data; the image encompasses the same region in central Vietnam.*

5.2.2. Scales and Spatial Resolutions

Spatial resolution, which is the size of an area on the ground expressed in a pixel or the ground sampled distance, is another crucial feature of remotely sensed imaging. Typically, the resolution of an image is indicated by a single number, like 30 m. This number corresponds to the length of one side of a square pixel when projected onto the surface of the Earth. The length and breadth would be supplied if the pixel were rectangular, which is not a very frequent shape (Marceau et al., 1994).

So, how does one determine the most suitable resolution for a specific project? As a general guideline, the resolution ought to be roughly 10 times smaller than the feature size being analyzed. For instance, if one aims to identify patches of forest measuring 1 ha (10,000 m^2) in size, a pixel resolution of around 32 m (¼1024 m^2) would be ideal. Regrettably, in reality, these general guidelines are not always feasible as various factors can impact the ability to accurately identify features on the surface of Earth. In cases where the landscape exhibits significant heterogeneity or low contrast between the features, a more precise resolution than that determined by applying the rule of thumb may be necessary to distinguish between distinct features (Atkinson & Aplin, 2004).

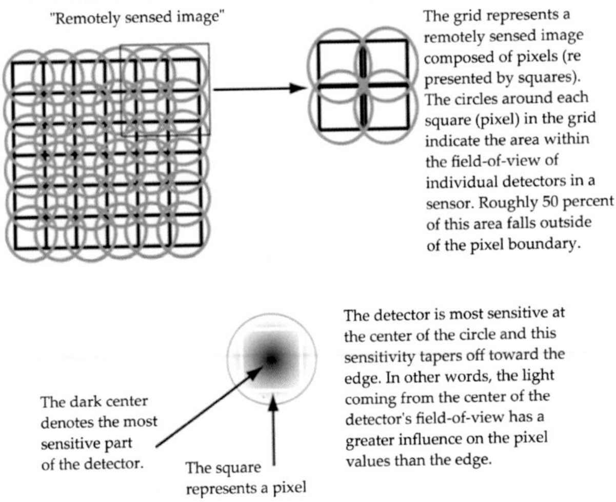

Figure 5.4. *The area captured by a single satellite-based sensor detector is depicted.*

The size of a feature that can be detected with a given resolution is greatly influenced by the contrast between surrounding cover types and the actual features. For instance, given the strong contrast between the bright dirt and the dark forest in imagery with 30 m pixels, a few meters wide trail, passing through a dense forest could be identified with a 30 m resolution imagery

(Figure 5.5). Put simply, the reflected light from the trail is sufficiently bright to distinguish it from the surrounding forest pixels. However, mapping different types of forests may pose a challenge due to the limited contrast between the various types (Qi & Wu, 1996). It is worth noting that the information within a pixel does not accurately represent a square on the ground. A sensor detector records the energy for a single pixel by viewing a circular or elliptical area on the ground. In Figure 5.4, it is evident that about half of the data captured by a pixel is derived from the surrounding surface area. This often leads to blurred boundaries between objects instead of the sharp edges one would anticipate (Treitz & Howarth, 2000).

Figure 5.5. *Landsat ETM+ image of the area surrounding Sakaraha in southern Madagascar (path 160, row 76, captured on August 23, 2000).*

The degree of landscape heterogeneity can also affect the ability to identify features. When the landscape is highly heterogeneous in terms of terrain or land cover, it becomes more challenging to identify individual features compared to a more homogeneous landscape.

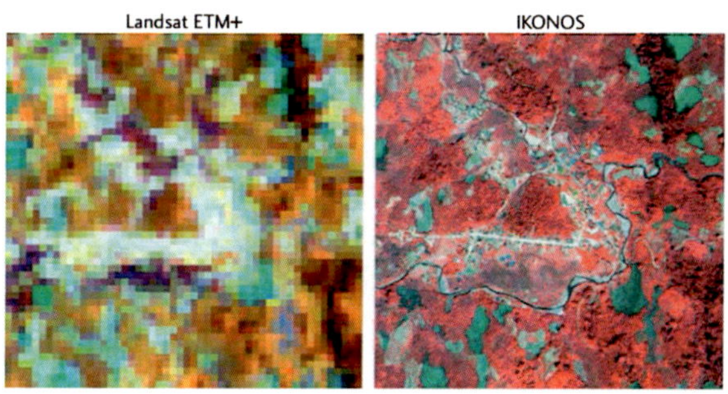

Figure 5.6. *Comparing the spatial resolutions of a 30 m spatial resolution Landsat ETMÍ with a 1 m spatial resolution IKONOS image.*

Source: MDPI; Creative Commons License.

Resolution greatly affects the sensitivity to heterogeneity (Hay et al., 1997). Certain landscapes may seem more heterogeneous when observed at fine resolutions but appear more homogeneous when observed at coarser resolutions (Figure 5.6). Hands-on experience is often the most effective way to grasp the constraints of a specific resolution. There are always exceptions to every rule of thumb. Seeking guidance from individuals with experience in a related project can provide valuable assistance (Raziei et al., 2013).

5.2.3. Temporal Options: Acquisition Frequency and Seasonality

When requesting or interpreting images, it is necessary to take the local weather and season into account. In certain areas, it is only possible to obtain images during seasons when there is minimal cloud cover. It is important to consider the condition of the features that need to be identified. For instance, when someone wants to map where a deciduous forest is, they may want to get pictures of the trees when they have leaves on them (Alexandridis et al., 2015).

For effective mapping of deciduous and evergreen vegetation, it would be beneficial to obtain imagery during both the leaf-on and leaf-off seasons. When obtaining several photographs of an area for tracking a specific phenomenon like land cover change, it is advisable to capture the images on "anniversary" dates.

This ensures that all the imagery is taken around the same time of the year, minimizing any changes that could be attributed to phenology. Typically, the higher the frequency at which a satellite sensor captures images of the area under investigation, the better the chances of finding a cloud-free image that meets the needs (Helfat & Eisenhardt, 2004).

5.2.4. Imagery Requisition

As the remote sensing industry has matured, the ease of ordering imagery has greatly improved. Most satellite vendors, private and government, provide user-friendly Websites to browse image archives for existing imagery and to place an order for a new image acquisition. Unfortunately, understanding the details of the various options can still be somewhat confusing, and it can require significant effort before the Web interfaces tell you how much your request might cost (Cohen, 2007).

When ordering imagery, there are various questions the user needs to answer about the product's □ualities. These in□uiries deal with the image file format, the kind of geometric processing (transforming the image to fit a specific orientation or map projection), and radiometric processing (changing the pixel values to represent physical values like radiance or reflectance).

Basic geometric and radiometric processing involves utilizing data captured by satellites or ground-tracking stations. Because it is automated, this kind of processing is reasonably priced. If a higher level of accuracy is required, ancillary data must be utilized. For example, a digital elevation model can be utilized to address geometric distortions resulting from changes in terrain or to standardize image pixel values by minimizing the impact of shadows and highly illuminated slopes. This advanced processing level can greatly impact the cost of the imagery. With access to the right software and skilled image processing personnel, it is feasible to perform customized processing of the imagery internally. While considering the processing type to request, it is important to have a clear understanding of the level of precision and accuracy needed for the final product (Schuster & Gale, 1996). For instance, when it comes to geometric processing, is it considered acceptable for the location error of individual pixels to be in the range of several 100 meters, dozens of meters, or less than 10 m?

While it is beneficial to have access to the processing options provided by image vendors, it can be helpful to seek clarification on any options that may be unclear. The cost of satellite imagery can vary greatly, from being available for free to exceeding US $100 per square kilometer. Two key factors that have a direct impact on the cost of imagery are the processing levels required and the spatial resolution. The finer the resolution, the higher the cost of the imagery. Similarly, the more processing that is required, the more expensive the imagery becomes. For any requests that involve programming the sensor to capture a new image, a sizable fee might be associated. Occasionally, the costs may vary depending on the urgency of the requests, with faster turnarounds incurring higher charges (Kahn et al., 1989). One aspect to consider when it comes to purchasing satellite imagery is the licensing agreement. Often, when someone purchases an image, they acquire a license to use it rather than owning the image outright. It is crucial to have a clear understanding of the license agreements as they outline the guidelines for sharing and publishing imagery, as well as any products derived from it (Gorelik & Gyftopoulos, 2021).

5.2.5. Making the Decision

So, how does one navigate through the various options of image distributors and imagery to choose the most suitable ones? Finding the most suitable imagery for a specific project can be quite a challenge. The following sections of this book aim to equip you with enough knowledge to make informed decisions when the ideal imagery is not available. Anyone who starts utilizing satellite imagery, soon realizes that the perfect scenario is rare and careful consideration is required to make the most of the available resources (Milkman et al., 2009).

Discussing with people who have completed related work is always a smart move. Professional societies can provide useful information and several list servers and Internet newsgroups that study remote sensing and can be excellent places to ask questions about choosing suitable pictures (Saaty, 1990).

5.3. SOFTWARE OPTIONS

If the image data is in a digital format, computer software is necessary to visualize those images. There are numerous software options to choose from when it comes to viewing RS imagery. This section provides an overview of various software categories, such as specialized remote sensing software, GIS, and common graphics software (Taudes, 1998).

5.3.1. Graphics Software

With the growth of digital photography, graphics software has become the most prevalent form of image-processing software. Regrettably, not all software designed for digital photography processing is suitable for satellite imagery processing. One issue is that digital photo processing usually deals with color images composed of three "bands" or "layers" (red, green, and blue). However, RS or other satellite data can have multiple bands, which can complicate the processing. The majority of entry-level graphics software programs are limited to working with three-band (color) or single-band (black and white) (McNicholas et al., 2011). Advanced graphics software, such as Adobe Photoshop, offers the ability to manipulate image bands, making it a valuable tool for enhancing and visualizing satellite imagery. However, one limitation is that graphics software is unable to utilize geocoding information from satellite images. This can pose challenges when accurately overlaying vector data such as contours, towns, and roads, or when merging multiple images into a single mosaic. Software like the GNU Image Manipulation Program (GIMP) or Photoshop can be a great starting point for exploring satellite imagery (Delp & Loan, 1995).

5.3.2. GIS Software

GIS software is another tool that can be utilized to view and, in certain instances, analyze remotely sensed data. Some GIS packages offer restricted abilities to display image data, and cheaper options often lack advanced graphic enhancement and processing features (Wang, 2014). With the increasing overlap between GIS and remote sensing, the functionalities of GIS and dedicated RS software have merged. One of the primary applications of satellite imagery in GIS is to serve as a background for various data layers, including roads, digital elevation, and political boundaries data. Using this tool can provide a highly effective way to showcase the land cover of a specific area, offering viewers a realistic depiction of the terrain (Steiniger & Hunter, 2013).

5.3.3. Specialized Remote Sensing Software

Specialized software for analyzing satellite imagery has integrated geo-coding features, but it can be costly and challenging to navigate. This software is utilized by professionals in the field of remote sensing and offers a wide range of advanced tools for visualizing imagery. Additionally, it provides an extensive set of image-processing tools for analysts to utilize. Specialized software is necessary for various types of processing, such as radiometric and geometric corrections, and automated land cover classifications (Raup et al., 2007). To effectively utilize these tools, a solid understanding of remote sensing is necessary. Many freeware and open-source software applications that are very simple offer a decent starting point for novice users with restricted features. Extensive experience is necessary to effectively utilize more advanced open-source and proprietary software. Commercial software vendors and organizations supporting open-source software offer sessions to educate users on various RS techniques (Marcus & Fonstad, 2010).

5.3.4. Mathematical and Visualization Software

Most of the time, mathematical and visualization software is used for analyzing the numbers that make up the images. This tool is designed for advanced users who have a deep understanding of computer programming and image processing. Within research settings, this software is utilized to develop specialized programs for analyzing and visualizing imagery in ways that far surpass the capabilities of standard software routines (Asmuss & Budkina, 2019).

5.3.5. Deciding Which Software to Use

When it comes to choosing software, several factors come into play, including budget, level of expertise, and the specific tasks at hand. Photoshop or a comparable program might be a fantastic place to start for someone who is familiar with it and is just interested in dealing with visual pictures. You should use a GIS package that would be better suited for someone who wants to combine RS images with other data, like vector GIS layers, or if they want geospatial features, like being able to use geographic coordinates such as latitude and longitude to find a feature, but don't need to do a lot of image analysis. For those interested in advanced image-processing capabilities and an openness to learning more sophisticated techniques, it could be beneficial to consider acquiring specialized remote sensing software (Lai et al., 2002).

5.4. VISUALIZING THE DATA

Depending on the software, there are certain approaches to viewing satellite imagery. GIS and other RS software typically offer a range of tools to manipulate and visualize satellite imagery. Let us delve into the essential concepts required to manipulate remotely sensed imagery and create valuable image products for interpretation or display (Olshannikova et al., 2015).

5.4.1. Creating a Color Composite by Combining Bands

Assigning distinct satellite image bands to the RGB colors emitted by the computer monitor allows the alteration of the appearance of the image when creating color composites of satellite image bands (Figure 5.7). Various band combinations can be utilized to enhance the visualization of specific phenomena. The band assignment used to create each image is provided below. These figures represent some of the commonly used band combinations. However, when working with imagery, there is room for experimentation with different band combinations (Gillespie et al., 1987).

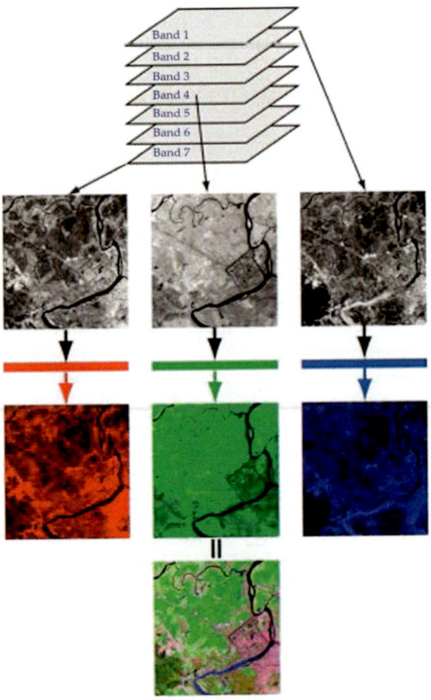

Figure 5.7. *A color rendering of a multispectral image obtained by projecting red, green, or blue light onto three bands selected from the image.*

Source: MDPI; Creative Commons License.

5.4.2. Image Enhancements

Upon initial display, satellite images may appear dark and lacking in contrast. When the image appears visually appealing upon opening, it is likely that the image has been enhanced beforehand or that the software has an automatic enhancement feature. The image enhancement tools are included with the GIS or image processing program to brighten the image and boost contrast so that the objects of interest can be identified more easily if there is a need to improve the image's appearance (Ortiz et al., 2012).

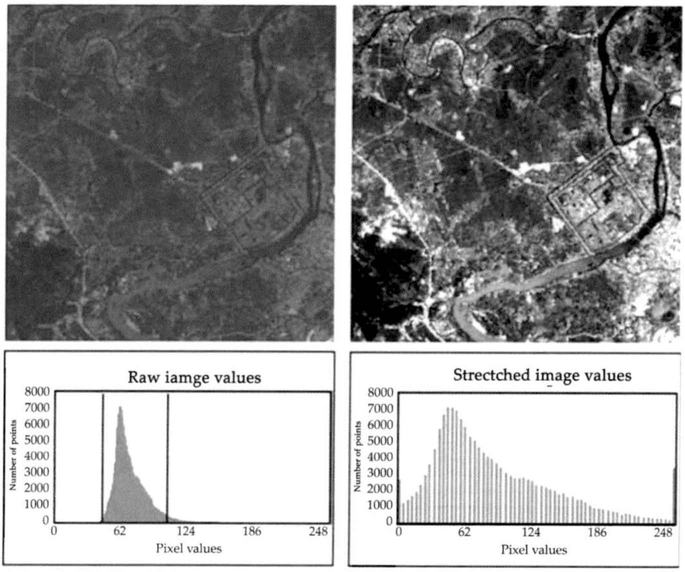

Figure 5.8. *The image on the left shows the raw Landsat ETM+ band 3 (red, 630–690 nm) image, with the pixel value histogram displayed below. The histogram indicates that pixel values range from 42 to 220. However, most values fall between approximately 45 and 115, leaving a portion of the data range unused. Pixel values between 0 and 48 were set to 0, and values between 111 and 255 were set to 255 to brighten the image on the right. Values from 48 to 111 were linearly mapped to a range of 0 to 255, as shown in the histogram on the right.*

Before delving into the topic of image enhancement, it is crucial to understand the constraints and objectives that must be considered. Computer monitors can display a wide range of pixel values, typically from 0 to 255. The value '0' corresponds to no light, or black, while '255' represents the maximum intensity of light or white. One common issue with satellite imagery is the concentration of pixel values within a limited data range (Figure 5.8). To utilize the entire range of available display values (often 0–255; Figure 5.8), these image values can be redistributed or scaled (Aghagolzadeh & Ersoy, 1992).

While each sensor has a different potential data range, the two most commonly used ranges are 0–1023 and 0–255. It is possible to adjust the initial image values to ensure they can be properly viewed on a computer screen, regardless of the data range.

Typically, image enhancement aims to enhance the clarity of an image, enabling us to more easily identify the desired features. This is achieved by adjusting the histogram to fully utilize the range of values from 0 to 255. The data is then partitioned inside the histogram to enhance visual interpretation (Koo & Cha, 2017).

An image's appearance can be enhanced in a variety of ways. One commonly used method involves adjusting the contrast and brightness of the image. Adjusting these two aspects is easy to understand, but it has some limitations in terms of capabilities. The linear stretch is a commonly used method among many. As shown in Figure 5.8, a linear stretch trims the tails of the histogram, and the dynamic range is expanded by stretching the rest of the histogram across the entire 0–255 data range. The pixel values in the lower part of the histogram are set to a value of 0, while the values in the upper part of the histogram that exceed the range are set to 255.

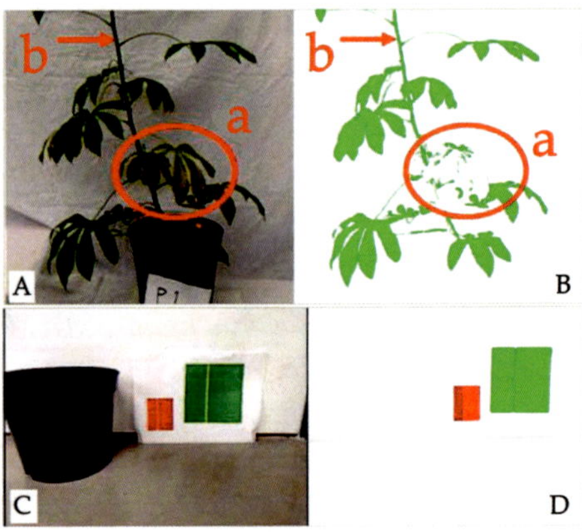

Figure 5.9. *Raw (A) and processed (B) images of the cassava plant (Manihot esculenta Crantz) with the raw known (C) and selected (D) red and green calibration areas. The areas colored green (□) are identified as leaves, □ut the yellow/senesced leaves (a) are omitted by the easy leaf area software.*

Source: Wasonga, D. O., Yaw, A., Kleemola, J., Alakukku, L., & Mäkelä, P. S. A. (2021). Red-green-blue and multispectral imaging as potential tools for estimating the growth and nutritional performance of cassava under deficit irrigation and potassium fertigation. Remote Sensing, 13(4), 598. https://doi.org/10.3390/rs13040598 (Creative Commons License).

Many of the software programs offer the option to input a variable that determines the extent of the clipping of the histogram tail. It can be expressed as the number of standard deviations, or 2 SD, from the mean, or as a percentage of the histogram, or 2%, that needs to be clipped. Various techniques can be employed to enhance the visual appeal of an image, aside from adjusting brightness and contrast settings (Lee, 1980). Several techniques can be used, such as Gaussian stretches, histogram equalization, and linear stretches among others (Figure 5.9).

Some image-processing software offers a feature that allows defining the specific area within the image to select pixels for creating the histogram. It is an incredibly valuable feature as it enables the maximization of the potential of different features, like bodies of water or forests. In Figure 5.10, the image demonstrates the distinct impact of stretching an image via either the histogram of the entire image or the histogram solely focused on the water portion. In the second image, there are noticeable variations in the water not apparent in the first image (Gharbi et al., 2017). Certainly, the second image appears to have a washed-out appearance overall. However, if the focus is on the water as the main point of interest, this enhancement reveals valuable information that is not present in the former image.

Figure 5.10. *Normal RGB image (A); and (B) the green area of cassava plant (Manihot esculenta Crantz) identified as leaves □y the easy leaf area software after thresholding.*

Source: Wasonga, D. O., Yaw, A., Kleemola, J., Alakukku, L., & Mäkelä, P. S. A. (2021). Red-green-blue and multispectral imaging as potential tools for estimating the growth and nutritional performance of cassava under deficit irrigation and potassium fertigation. Remote Sensing, 13(4), 598. https://doi.org/10.3390/rs13040598.

There might be a few more ways to stretch the data to enhance the imagery, subject to the software being used. One can gain a better understanding of the various enhancement options by conducting experiments with different methods to determine the most effective ones.

Certain software packages offer a visual representation of the histograms before and following enhancement. By studying these histograms, one can gain a solid understanding of the basic changes that take place during image enhancement (Rahman et al., 2004).

5.4.3. Using a Panchromatic Band to Sharpen Image

"Image sharpening" or "image fusion" is another type of image enhancement that combines two different types of multispectral images—one with a higher resolution (typically a single-band panchromatic image) and the other with a lower resolution—to produce a multispectral image with greater detail than the initial multispectral image.

There are multiple choices for panchromatic or pan-sharpening in a wide range of RS image-processing software packages. While the output image may contain more spatial detail, it does not possess the same qualities as if it were initially captured at the panchromatic image's higher resolution (Vijayaraj et al., 2006) (Tables 5.1 and 5.2).

Table 5.1. *Summary of Different Types of Landsat Images Indicating Spatial, Temporal, Radiometric, and Spectral Resolution*

Landsat 1–3 (MSS) [1] 1972–1983			Landsat 4–5 (MSS) 1975–2013			Landsat 4–5 (TM) 1975–2013			Landsat 7 (ETM+) 1999 to Present			Landsat 8 (OLI) 2013 to Present		
Temporal 18 days	Radiometric 6 bits		Temporal 18 days	Radiometric 6 bits		Temporal 16 days	Radiometric 8 bits		Temporal 16 days	Radiometric 9 bits		Temporal 16 days	Radiometric 12 bits	
Band Name	Spectral (µm)	Spatial (m)	Band Name	Spectral (µm)	Spatial (m)	Band Name	Spectral (µm)	Spatial (m)	Band Name	Spectral (µm)	Spatial (m)	Band Name	Spectral (µm)	Spatial (m)
Band 4-Green	0.5–0.6	60	Band 4-Green	0.5–0.6	60	Band 1-Blue	0.45–0.52	30	Band 1-Blue	0.45–0.52	30	Band 1-Ultra	0.43–0.45	30
Band 5-Red	0.6–0.7	60	Band 5-Red	0.6–0.7	60	Band 2-Green	0.52–0.60	30	Band 2-Green	0.52–0.60	30	Band 2-Blue	0.45–0.51	30
Band 6-NIR	0.7–0.8	60	Band 6-NIR	0.7–0.8	60	Band 3-Red	0.63–0.69	30	Band 3-Red	0.63–0.69	30	Band 3-Green	0.53–0.59	30
Band 7-NIR	0.8–1.10	60	Band 7-NIR	0.8–1.10	60	Band 4-NIR	0.76–0.90	30	Band 4-NIR	0.77–0.90	30	Band 4-Red	0.64–0.67	30
												Band 5-NIR	0.85–0.88	30
						Band 5-SWIR1	1.55–1.75	30	Band 5-SWIR1	1.55–1.75	30	Band 6-SWIR1	1.57–1.65	30
						Band 7-SWIR2	2.08–2.35	30	Band 7-SWIR2	2.09–2.35	30	Band 7-SWIR2	2.11–2.29	30
									Band 8-Pan	0.52–0.90	15	Band 8-Pan	0.50–0.68	15
												Band 9-Circus	1.36–1.38	30
						Band 6-TIR	10.40–12.50	120	Band 6-1-TIR	10.40–12.50	60	Band 10-TIR	10.60–11.19	100
									Band 6-2-TIR	10.40–12.50	60	Band 11-TIR	11.50–12.51	100

Source: Phiri, D., & Morgenroth, J. (2017). Developments in Landsat land cover classification methods: A review. Remote Sensing, 9(9), 967. https://doi.org/10.3390/rs9090967.

* The original pixel size for Landsat MSS was 79 × 57 m; however, most literature reports a spatial resolution of 60 m because the data has been resampled to a 60 m pixel size.

Table 5.2. *Summary of Land Cover Classification Overall Accuracies with Different Classifiers Applied on Specific Landsat Images and Land Cover Type*

Classification Approach	Method	Classifier Used	Landsat Images Used	Type of Land Cover	Accuracy Attained (%)	References
Pixel-based	Supervised	ML, NN, SVM	MSS, TM, OLI	Urban area	73–82	[Poursanidis, D, Toll, D.L, Pullanikkatil, D et al.]
		ML	MSS, TM, OLI	Forest plantation	61–90	[Pullanikkatil, D.; Sloan, S., Kumar, R.; et al.]
		ML	MSS, OLI	Dense forest	68–90	[Pullanikkatil, D – Kumar, R. et.al]
		ML	TM, OLI	Open forest	52–81	[Pullanikkatil, D, Vieira, I.C.G. et.al]
	Unsupervised	ISODAT	TM	Urban area	78–94	[Lunetta, R.S.; Justice, C.;et.al]
		ISODAT	TM	Forest plantation	71–87	[Justice, C.;– Lunetta, R.S.;et.al]
		ISODAT	TM, OLI	Dense forest	71–87	[Justice, C.;– Lunetta, R.S.et.al]
		ISODAT	TM	Open forest	69–81	[Justice, C.;– Lunetta, R.S.et.al]
	Contextual	ECHO, majority filter	TM	Urban area	72–81	[Stuckens, J.; Flygare, A. et.al]
		ECHO, majority filter	TM	Forest plantation	70–81	[Stuckens, J.; Flygare, A. et.al]
		ECHO, majority filter	TM, ETM+	Dense forest	72–82	[Stuckens, J.; Flygare, A. et.al]
		NN	MSS	Open forest	66–90	[Swain, P.H.; Stuckens, J.; Flygare, A. et.al]
		ECHO, majority filter	TM, ETM+	Agricultural area	66–97	[Stuckens, J.; Flygare, A. et.al]
	Hybrid	ISODAT, fuzzy, ML	TM, ETM+	Urban area	64–96	[Lo, C.; Kuemmerle, T.; et.al]
		ML, rule-based, ISODAT	TM, ETM+, DEM	Forest plantation	74–87	[Lo, C.; Kuemmerle, T.; et.al]
		ML, rule-based, ISODAT	TM, ETM+, DEM	Dense forest	79–91	[Lo, C.; Kuemmerle, T.; et.al]
		ML, rule-based, ISODAT	TM, ETM+	Agricultural area	64–84	[Lo, C.; Kuemmerle, T.; et.al]

Sub-pixel	SMA	LSMA, MESMA	TM, ETM+, OLI	Urban area	83–90	[Poursanidis, D.; Powell, R.L.;et.al]
		LSMA	TM	Forest plantation	77–93	[Peterson, S.H.; Roberts, D.A.; et.al]
		LSMA	TM, OLI	Dense forest	75–93	[Mayes, M.T.; Peterson, S.H.;et.al]
		LSMA	TM	Open forest	77–87	[Peterson, S.H., Hamada, Y.;et.al]
		LSMA	TM, OLI	Agriculture area	70–74	[Mayes, M.T.; Théau, J.;et.al]
	Fuzzy analysis	Fuzzy C-mean	MSS	Urban area	70–90	[Fisher, P.F.; Melgani, F.; et.al]
		Fuzzy partitioning	TM	Forest plantation	74–90	[Mota, G.L.A.; Wang, L.et.al]
		Fuzzy membership	TM	Dense forest	74–70	[Fisher, P.F.; Foody, G.; et. al]
		Explicit fuzzy	TM	Open forest	56–79	[Fisher, P.F.; Melgani, F.; et.al]
		Explicit fuzzy	TM	Agriculture	74–92	[Mota, G.L.A.; Melgani, F.; et.al]
Object-based	OBIA*	SVM, DT, RF, NN	ETM+, TM, MSS, OLI	Urban areas	73–98	[Poursanidis, D.; Samal, D.R.; et.al]
		Decision rule	ETM+, TM	Forest plantation	80–97	[Newman, M.E.; Samal, D.R.; Dingle Robertson, L.; et.al]
		Decision rule	TM	Natural forest	77–95	[Newman, M.E.; Dorren, L.K.A.; Samal, D.R.; Dingle Robertson, L.; et.al]
		Decision rule	TM	Agriculture area	76–90	[Dorren, L.K.A.; Dingle Robertson, L.; et.al]
	Knowledge-based	Expert-knowledge	MSS, TM	Urban area	87–90	[Tailor, A.; Wang, L.; Ton, J.; et.al]
		Spectral expert	MSS, TM, DEM	Forest plantation	86–94	[Ton, J.; Manandhar, R.; et.al]
		Spectral expert	MSS, TM, DEM	Dense forest	85–92	[Ton, J.; Manandhar, R.; et.al]
		Eco-SDSS	MSS, TM, GIS	Agriculture area	85–88	[Sikder, I.U.; Ton, J.; et.al]

Source: Phiri, D., & Morgenroth, J. (2017). Developments in Landsat land cover classification methods: A review. Remote Sensing, 9(9), 967. https://doi.org/10.3390/rs9090967.

* The scale parameter range between 5 and 10 for Landsat MSS and 5–20 for Landsat TM, ETM+, and OLI while shape and compaction were reported to be 0.1–0.5 and 0.5–0.8, respectively.

Note: the values of accuracies presented range from the lowest to the highest overall accuracies for each land cover type. [ML: maximum likelihood; NN: nearest neighbor; LSMA: linear spectral mixture analysis; MESMA: multiple endmember spectral mixture analysis; and RF: random forests].

The panchromatic image with higher resolution can be obtained from various sources, although it is commonly derived from the same sensor that captured the multispectral data. As an illustration, the Landsat ETMþ sensor features a panchromatic band with a resolution of 15 m and can capture multispectral imagery at a 30 m resolution. The panchromatic band is made up of information spanning all colors or image bands. One of the benefits of such a system is that it captures images at the same time, ensuring that they both cover the same areas on the earth (King & Wang, 2001).

5.4.4. 3D Point of View

By examining a satellite image of the Earth, one can easily distinguish between flat and mountainous regions. It is primarily a result of the sun's illumination, the positioning of shadows, and how human brains interpret this data. Dealing with shadows in image classification can be quite challenging, yet they can offer valuable insights into the terrain and serve as useful indicators for identifying ground features. There are instances when our minds deceive us, causing us to perceive valleys as mountains and mountains as valleys when we observe an image (Dhome et al., 1989). Creating a 3D perspective image is a convenient method to accentuate or enhance the visualization of terrain relief. This perspective is achieved by overlaying an image onto a digital elevation model or DEM. Specific software is necessary to achieve this effect, but luckily there are free options available. Additionally, numerous software packages for remote sensing image processing offer this capability (Banta et al., 2000). To construct the views, perspective data is supplied, including the location of the sun (azimuth and elevation), the viewer (elevation, longitude, or latitude), and the exaggeration of the topography to highlight the relief of the terrain. Certain software enables one to soar above the terrain as if they were piloting an aircraft. In addition, satellite images can be overlaid onto a land cover or any other type of map. Working with these three-dimensional tools is not only enjoyable, but it also offers valuable insights into satellite images that are not readily apparent in standard two-dimensional images (Kwon et al., 2020).

5.5. FURTHER PROCESSING

There are instances where people can begin utilizing satellite imagery immediately for their specific tasks. In other cases, it may be necessary to process the imagery to refine its radiometric or geometric properties. This type of processing, known as preprocessing, can vary in complexity. This section will provide an overview of these processing techniques and discuss the challenges associated with each type of processing (Cramer, 2000) (Table 5.3).

Table 5.3. *Different Strategies and Methods for Preprocessing Steps of Landsat Change Detection Studies*

	Strategy	**Method**
Geometric	Automated	L1TP standard data
	User-based	Coregistration
		Polynomial fitting
		Commercial software
Atmospheric	Automated	Surface reflectance data
		LEDAPS
		LaSRC
		L8SR
	User-based	MODTRAN
		LOWTRAN
		FLAASH
		DOS
		COST
		ATCOR
		6S
Cloud/Cloud Shadow	Automated	Fmask
		CFmask
		Tmask
		Cloud score
		QA band
	User-based	Manual mask
		Thresholding
		Cloud-free image selection

Source: Hemati, M., Hasanlou, M., Mahdianpari, M., & Mohammadimanesh, F. (2021). A systematic review of Landsat data for change detection applications: 50 years of monitoring the Earth. Remote Sensing, 13(15), 2869. https://doi.org/10.3390/rs13152869.

5.5.1. Radiometric Processing

The primary objective of radiometric processing is to accurately calibrate the pixel values captured by the sensor. This calibration ensures that the pixel values correspond to specific physical properties, like reflectance or radiance. Additionally, radiometric processing can be used to normalize multiple images, ensuring that similar features have consistent pixel values across all the images. To grasp the process, it is essential to comprehend the perspective of a satellite sensor as it captures an image (Kashani et al., 2015).

There is a common misconception that satellite sensors record reflectance coming from features on ground level. However, a satellite sensor measures energy or radiance at the sensor for a specific range of wavelengths. Understanding radiance is crucial for accurately interpreting sensor data. The upcoming section will explore the connection between surface reflectance and radiance at the sensor, which may not be as robust as desired (Schwarz et al., 1994).

Satellite sensors are usually calibrated before launch to ensure accurate measurement of radiance. However, over time, recalibration becomes necessary to account for changes and degradation of the sensor. Luckily, the majority of satellite sensors can self-calibrate. Radiometrically calibrated satellite images are what a customer gets when they order them. They have been corrected using on-board calibration data to give sensor radiance. The image received on the user end, on the other hand, will not contain the radiance values within the image (Alonso et al., 2014). Instead, the image is typically scaled from 0–1023 or 0–255. Typically, the necessary information to calculate the radiance values can be found in the image metadata. These data can be processed for each band using remote sensing image-processing software. This allows you to calculate the actual radiance values, frequently supplied as maximum and minimum or offset, and gain values radiance values. When attempting to calculate reflectance values or compare the pixel values (digital numbers) from two photos, this step is required. Calculating the at-sensor radiance is not necessary if all you are interested in is a visual image; the image will appear the same whether physical radiance is calculated or not (Honkavaara et al., 2009).

5.5.2. Reflectance – Remote Sensing "Holy Grail"

Surface reflectance refers to the percentage of the light reflected from a specific feature on the Earth's surface compared to the amount of light that falls on that feature. Accurate calculation of surface reflectance is widely regarded as the holy grail of remote sensing and can be a challenging endeavor. It would be incredibly convenient if satellite instruments would effortlessly measure surface reflectance. This would greatly simplify the analyst's work and result in significantly more precise land cover maps (Ballatore et al., 2014).

Land cover mapping and RS interpretation in general heavily depend on the assumption that Earth's features possess distinct spectral signatures. The spectral signature of a feature can be represented by a spectral reflectance curve, which shows the reflectance values of that feature throughout the EM spectrum. Using specialized software, an image analyst's task is to classify image pixels into relevant land cover groups according to pixel values, which are fre□uently taken to be directly linked to the reflectance of the feature. For instance, every pixel with a value comparable to a pixel that we know is water would be classified as such, and every pixel with a value comparable to a pixel that we know is forest will be classified as such, and so on, until the entire image is assigned a particular land cover type. Although this seems quite simple, it can be highly challenging because it is difficult to accurately measure the reflectance of a ground-based surface feature from a satellite image (Shabala, 2003).

To gain a better understanding of this issue, we can compare satellite remote sensing with reflectance measurements can be compared with satellite remote sensing in a lab. The following are important things that affect reflectance:

i. The incoming radiation intensity: refers to the amount of light energy that is received by the target.

ii. The reflected radiation intensity: refers to the strength of the light energy immediately after it has left the target.

iii. Detector and light source position in relation to the target (Cavender et al., 2022).

Determining a material's reflectance □ualities in a lab setup using specialized equipment is simple. Measuring the intensity of light energy is a straightforward task. There is a short light path, and the quality of air is controlled, minimizing energy loss between the light source, target, and detector. Additionally, the target orientation and illumination can be determined with great accuracy and precision.

When it comes to satellite remote sensing, the situation is quite distinct. Understanding the behavior of light as it travels through the Earth's atmosphere is crucial. The interaction between light and various particulates such as smoke, dust, and water vapors can have a profound impact on the signal before and after it reaches the target. While a lot of work has been done in the removal of these atmospheric effects in photos, they are still not easy or consistent to get rid of. Finding out the benthic surface reflectance in a small body of water is even harder because of these kinds of effects. Next, it is necessary to eliminate the impact of atmospheric conditions and the presence of water on the surface and in the columns (Lausch et al., 2016). Regarding illumination and detector orientation, it is possible to determine the position of the satellite and the sun at the time the image was captured. However, determining the orientation of the target, such as its aspect and slope, is more challenging. Although digital elevation models (DEMs) typically provide only a rough approximation of the target's orientation, they can be used to estimate these characteristics. For example, the same type of land cover might be found in two different locations within an image: one under a mountain's shadow and another positioned where the sensor receives the maximum amount of reflected energy. Processing these two locations to have identical reflectance values can be □uite difficult (Schaaf et al., 2011).

It is important to note that land cover features can vary in appearance throughout one satellite image. There can also be some confusion when distinguishing among land cover classes that have similar reflectance signatures, like various types of forests. This poses difficulties for categorization and interpretation techniques that are both automated and visual.

5.5.3. Geometric Processing

Images that have undergone geometric correction can be compared to other images (image-to-image registration) or maps (image-to-map registration). Various geometric correction levels can be used on satellite images to rectify inherent geometric errors. One of the simplest uses systematic corrections to get rid of the distortions that come from the sensor and the satellite's movement around the Earth. The method commonly utilizes ephemeris data, which is a record of the satellite's velocity and location during image acquisition. This data is used to determine the position in real-world geographic coordinates, that is, latitude, longitude, or projected coordinates like the Universal Transverse Mercator (UTM). In regions with limited relief in the landscape, the generated image frequently has a positional displacement of less than 50 meters, indicating a reasonable level of accuracy (Toutin, 2004).

Image registration, an intermediate process, can be used to enhance the positional accuracy of satellite images. One way to improve the precision of an image's coordinate system is by utilizing ground control points (GCPs). GCPs are identifiable locations with known coordinates in both the real world and the image. Using various algorithms, the image is stretched or "warped" to achieve the optimal fit of the GCPs and the satellite image. These algorithms mimic the traditional technique of "rubber-sheeting," where an aerial photo is printed on a rubber sheet and then stretched to align with a map reference. Unfortunately, in regions with significant terrain variations, this approach often yields subpar results unless a substantial number of GCPs are employed

(Pottmann et al., 2009). A more advanced method for geometric correction is known as "orthorectification." This method uses elevation information, typically obtained from a digital elevation model (DEM), to correct distortions caused by variations in terrain. Orthorectification is particularly useful in areas with mountainous topography. An orthorectified image can be compared to a topographic map in terms of its spatial accuracy, serving as an image map with precise spatial information. When dealing with flat terrain, basic image registration techni□ues yield results comparable to an orthorectified image, as elevation changes do not need to be considered (Tasdizen et al., 2003).

Orthorectified images are far superior to non-orthorectified ones due to their exceptional absolute positional accuracy. There can be a problem that arises if comparing orthorectified and non-orthorectified images. When comparing two images to track shifts in land cover as time passes, the alignment between the images is crucial. The relative positional accuracy, which measures how well the images are aligned to each other, takes precedence over the absolute positional accuracy, which focuses on how accurately the image and map base match. The intricate distortion adjustments in an orthorectified image make it extremely difficult and fre□uently impractical to try to match an image that has been geocorrected using basic image registration techniques (Braun & Zhao, 2005).

Rather than using image-to-map registration, a method known as image-to-image registration can be used to match two photographs covering the same area. This process involves selecting one image as a reference and aligning the other image(s) to it. Control points that are visible in both images must be selected to complete the process. Once these control points are identified, the same algorithms used in image-to-map geo-referencing are applied. In many cases, registering two images that have undergone similar geometric processing makes it easier to achieve high relative accuracy between multiple satellite photos. Essentially, if one image has been systematically corrected, it is generally simpler to register it with another systematically corrected image rather than with an orthorectified image (Besl, 1990).

Depending on the level of accuracy, it might not be feasible to attempt to align multiple orthorectified images that were processed using different ortho-correction methods. Using different methods can result in varying distortions between both images, which can complicate the coregistration process. Ideally, it would be great to have all the images orthorectified via the same method for maximum accuracy. However, in real-world situations where financial and time resources are limited, this may not always be feasible (Elsey & Esedo, 2009).

In terms of accuracy, orthorectified images tend to have higher absolute precision. However, if relative accuracy is the main concern, it might be more beneficial to solely rely on systematically corrected images instead of combining them with ortho-imagery. Often, what functions best in a certain context will be determined by trial and error (Hu et al., 2021).

5.6. CLASSIFICATION AND INTERPRETATION OF IMAGES

After obtaining a satellite image and being able to see it on a computer screen, you undoubtedly want to be able to understand it. The main objective of the interpretation of an image is to recognize and classify features within an image. These features can encompass a wide range of objects or phenomena, including land cover, soil moisture, vehicles, urban features, or any other identifiable elements within the image (Kallipolitis et al., 2021).

Both automatic and visual methods can be used to identify features. By utilizing visual methods, one can easily connect the features observed in the image with familiar objects on the ground. By utilizing visual cues in the image, one can determine the identity of a specific feature. These cues include factors such as shadows, absolute and relative locations, patterns, textures, shapes, sizes, and colors. Although formal training is not necessary for visual interpretation, practice improves the ability to detect elements with more precision and accuracy (Santos et al., 2013).

There are several automated methods available for identifying features in a satellite image. Automated image classification is the most commonly used approach. Using this approach, a computer algorithm categorizes pixels in the image based on the classes defined by the analyst. The outcome is usually a map, like the land cover map. A different technique involves the use of automated feature recognition, where a computer algorithm is used to identify specific features as defined by the analyst. For instance, to find every house in a picture, an analyst could employ a feature recognition technique (Bernad et al., 2008).

5.6.1. Visual Interpretation

The methods used in the visual interpretation of satellite images closely resemble those developed more than a century ago for interpreting aerial photographs. Given the advancements in technology, visual interpretation is often undervalued despite its importance. Despite advancements in algorithms and computer speed, the human brain still outperforms machines in processing the extensive information needed for image feature identification (Pavlovic et al., 1997). There are several benefits to utilizing image photoproducts via visual interpretation:

- Creating a usable product takes less time.
- There is minimal to no additional cost involved apart from the initial image purchase.
- Brightly illuminated surfaces and shadows—lighting aspects that are typically difficult to interpret—can be employed as helpful tools.
- Basic remote sensing analysis skills are needed for image interpretation.
- Greater use of mental capacity.

1. **Less Time**: It may take months or years to finish a reputable land cover classification project. Visualization can begin as soon as the satellite data are obtained by using picture interpretation techniques. A computer with visualization software is all that is required to view the image on a screen, or it can be viewed on paper (Svatonova, 2016).
2. **Less Expense**: Classification projects, including land cover, can be highly expensive. The costs involve error analysis and the analyst's classification time. This typically involves conducting fieldwork to gather location and testing data, which can be quite costly, often reaching several thousand dollars. Typically, the expense associated with the data is only a small portion of the total cost of a land cover classification task (Mellet et al., 1996).
3. **Image Illumination**: An automatic classification system may encounter issues due to an unevenly illuminated satellite image. Accounting for these effects can be a considerable and occasionally fruitless endeavor. Visual interpretation, on the other hand, can utilize these challenges to assist in image interpretation. Understanding the impact of illumination variations on satellite images is crucial for identifying different features in the image (Fleming et al., 1991).
4. **Less Expertise**: Conducting an effective land cover classification re□uires significant expertise, and selecting the appropriate methods can be challenging. Interpreting an image using photo interpretation methods requires some training. However, the level of training needed to start interpreting an image is lower compared to the expertise needed for carrying out a land cover classification (Wang et al., 2015).
5. **More Brain Power**: A machine cannot compare to the human brain's ability to distinguish land cover elements in a satellite image. Computers excel at continually applying a particular set of rules to categorize an image. However, in practice, these rules are often not clear-cut or static (Borman & Grigg, 2009).

In addition, once a satellite image has been categorized, a thematic class is used to replace the visual cues present in the initial picture. Simply put, the image no longer shows the subtle changes throughout the forest; instead, it is replaced by one discrete class that represents a specific feature, like a forest.

Taking a map into the field—either in hardcopy or on a laptop or mobile device—is one of the finest methods to enhance your ability to interpret the information in an image. By properly locating yourself on the image and orienting it, you can view or visit a few of the areas depicted, as shown in Figures 5.1 and 5.2. Through the process of connecting visual observations with corresponding elements in the image, one can develop the ability to recognize and interpret various shapes and colors as real on-ground features. Finding comparable features in another location can start if you are able to recognize a few distinct features in the first area. This kind of experience is incredibly valuable and should not be underestimated. During their initial training, NASA astronauts engage in field exercises to hone their interpretation skills. These exercises involve analyzing astronaut-acquired and aerial photographs along with satellite imagery in real-world settings. This prepares them for the task of interpreting scenes they will encounter while in orbit (White et al., 1984). For better orientation on an image, one has the option to overlay features like towns, rivers, roads along other landmarks. You can use a GPS receiver to overlay latitudes and longitudes or any other coordinate grid on top of grid lines to get your exact location. If you happen to have a GPS and a computer, using a heads-up display or a moving map can be quite convenient. Using this setup, your location on the image is plotted automatically as you navigate the terrain, the display precisely follows your movements on the image, often using a cursor. One way to accomplish this is by utilizing GIS or software specifically designed for processing remote-sensing images (Kundel, 1990). When someone is proficient at correctly detecting image features, they can draw polygons on the image to delineate groups of pixels that represent specific features or types of land cover, thereby generating thematic maps, such as land cover maps. Essentially, they are using visual techniques to classify an image. This type of digitizing is commonly referred to as on-screen or heads-up digitizing, where the interpreter draws boundaries directly on the computer screen. This task can be quite time-consuming, particularly when working with large areas. However, the results can be just as effective as those achieved through automated methods (Hansen et al., 2011).

5.6.2. Classification of Images

Classified images, also known as maps, are commonly used for quantitative studies like GIS data analysis and modeling. Additionally, they are simple to understand because each class is represented by a distinct entity that is typically distinguished by color, making it easy to distinguish between them (Cheng & Chen, 2003).

Remote sensing literature has various comparisons of classification approaches and algorithms, with each one advocating for a specific method. While these studies hold value, it is important to consider the context in which they were conducted. For your application, you might not obtain similar accuracy using comparable techni□ues. Image classification continues to be a subject of ongoing research, with scientists working to enhance accuracy and offer guidance on selecting the most suitable approaches for various circumstances. To facilitate the discussion of various methods, they can be categorized into the following approaches (Vailaya et al., 1998).

5.6.2.1. Manual Classification

Classifying land cover manually, or visually, using remotely sensed data works well when the analyst is acquainted with the region being classed. This approach utilizes expertise in analyzing aerial photographs. The interpreter must use visual clues to distinguish between the various land cover classes, pattern, shape, texture, tone, and relationship to other objects. One of the main benefits of this method is its reliance on human intelligence to identify image features and connect them to real-world features. When it comes to accurately identifying image features, the human brain still has the upper hand over computers. One additional benefit is the ability to manually classify without relying on a computer, using a printed satellite image (Jeevanantham et al., 2015). One drawback of manual interpretation is its tendency to be time-consuming and laborious in comparison to automated classification. Additionally, since it relies only on a human interpreter, it is more prone to subjectivity. This limitation often hinders the application of this technique to smaller areas. One limitation is that it can only integrate RGB (three) bands of satellite image data, as the interpretation typically relies on a color image consisting of the said bands. Despite these limitations, manual classification remains a dependable approach for generating land cover maps.

Manual interpretation employs a rather straightforward technique. The analyst examines the image either on a printout or a screen and proceeds to outline specific land cover types by drawing polygons around them. By conducting land cover delineations on the computer screen, the land cover map is generated simultaneously with the delineation process. If a physical image is employed for interpretation, its final map will need to undergo digitization to be converted into a format that can be read by a machine (Passamonti & Maffioli, 2016).

5.6.2.2. Automated Classification

Most classification methods belong to this category. Using automated classification, analysts can utilize algorithms and the computational power of computers to create and implement specific rules for assigning pixels to different classes. While establishing these guidelines may involve some subjectivity, they are consistently implemented across the entire image, a task that visual methods cannot achieve. Typically, humans excel at utilizing location shape, and size details to recognize a feature in an image. However, computers outperform us when it comes to handling multiple layers of data, such as image bands and ancillary data (Caldas et al., 2002).

These algorithms have the capability to integrate data from numerous satellite bands and various ancillary data layers. In the past, automated algorithms were restricted to just using the pixel values within the image. However, there are now various approaches that enable users to effortlessly include additional data layers like soil type, aspect, slope, elevation, and other biophysical layers. This integration of ancillary data significantly enhances the accuracy of classification (Drăguț & Blaschke, 2006).

REFERENCES

1. Aghagolzadeh, S., & Ersoy, O. K. (1992). Transform image enhancement. *Optical Engineering, 31*(3), 614–626.
2. Alexandridis, T. K., Sotiropoulou, A. M., Bilas, G., Karapetsas, N., & Silleos, N. G. (2015). The effects of seasonality in estimating the C - factor of soil erosion studies. *Land Degradation & Development, 26*(6), 596–603.
3. Alonso, J., Batlles, F. J., López, G., & Ternero, A. (2014). Sky camera imagery processing based on a sky classification using radiometric data. *Energy, 68,* 599–608.
4. Asmuss, S., & Budkina, N. (2019). On usage of visualization tools in teaching mathematics at universities. *Engineering for Rural Development, 18,* 1962–1969.
5. Atkinson, P. M., & Aplin, P. (2004). Spatial variation in land cover and choice of spatial resolution for remote sensing. *International Journal of Remote Sensing, 25*(18), 3687–3702.
6. Ballatore, T. J., Bradt, S. R., Olaka, L., Cózar, A., & Loiselle, S. A. (2014). Remote sensing of African lakes: A review. In *Remote Sensing of the African Seas* (pp. 403–422).
7. Banta, J. E., Wong, L. R., Dumont, C., & Abidi, M. A. (2000). A next-best-view system for autonomous 3-D object reconstruction. *IEEE Transactions on Systems, Man, and Cybernetics-Part A: Systems and Humans, 30*(5), 589–598.
8. Bauer, M. E. (1975). The role of remote sensing in determining the distribution and yield of crops. *Advances in Agronomy, 27,* 271–304.
9. Bernad, G. P., Denise, L., & Refregier, P. (2008). Hierarchical feature-based classification approach for fast and user-interactive SAR image interpretation. *IEEE Geoscience and Remote Sensing Letters, 6*(1), 117–121.
10. Besl, P. J. (1990). Geometric signal processing. In *Analysis and Interpretation of Range Images* (pp. 141–205).
11. Borman, G. D., & Grigg, J. A. (2009). Visual and narrative interpretation. In *The Handbook of Research Synthesis and Meta-Analysis* (pp. 497–519).
12. Braun, W. J., Li, W., & Zhao, Y. Q. (2005). Properties of the geometric and related processes. *Naval Research Logistics (NRL), 52*(7), 607–616.
13. Brisco, B., Brown, R. J., Hirose, T., McNairn, H., & Staenz, K. (1998). Precision agriculture and the role of remote sensing: A review. *Canadian Journal of Remote Sensing, 24*(3), 315–327.

14. Caldas, C. H., Soibelman, L., & Han, J. (2002). Automated classification of construction project documents. *Journal of Computing in Civil Engineering, 16*(4), 234–243.
15. Cavender-Bares, J., Schneider, F. D., Santos, M. J., Armstrong, A., Carnaval, A., Dahlin, K. M., & Wilson, A. M. (2022). Integrating remote sensing with ecology and evolution to advance biodiversity conservation. *Nature Ecology & Evolution, 6*(5), 506–519.
16. Chang, C. T., Wu, Y. C., Lee, Y. K., & Chu, X. Y. (2018). Right metaphor, right place: Choosing a visual metaphor based on product type and consumer differences. *International Journal of Advertising, 37*(2), 309–336.
17. Cheng, Y. C., & Chen, S. Y. (2003). Image classification using color, texture, and regions. *Image and Vision Computing, 21*(9), 759–776.
18. Cohen, M. D. (2007). Accuracy of information on imaging requisitions: Does it matter? *Journal of the American College of Radiology, 4*(9), 617–621.
19. Cramer, P. (2000). Defense mechanisms in psychology today: Further processes for adaptation. *American Psychologist, 55*(6), 637.
20. Delp, S. L., & Loan, J. P. (1995). A graphics-based software system to develop and analyze models of musculoskeletal structures. *Computers in Biology and Medicine, 25*(1), 21–34.
21. Dhome, M., Richetin, M., Lapreste, J. T., & Rives, G. (1989). Determination of the attitude of 3D objects from a single perspective view. *IEEE Transactions on Pattern Analysis and Machine Intelligence, 11*(12), 1265–1278.
22. Dingle Robertson, L.; King, D.J. Comparison of pixel- and object-based classification in land cover change mapping. *Int. J. Remote Sens.* 2011, *32*, 1505–1529.
23. Dinguirard, M., & Slater, P. N. (1999). Calibration of space-multispectral imaging sensors: A review. *Remote Sensing of Environment, 68*(3), 194–205.
24. Dorren, L.K.A.; Maier, B.; Seijmonsbergen, A.C. Improved Landsat-based forest mapping in steep mountainous terrain using object-based classification. *For. Ecol. Manag.* 2003, *183*, 31–46.
25. Drăguț, L., & Blaschke, T. (2006). Automated classification of landform elements using object-based image analysis. *Geomorphology, 81*(3–4), 330–344.

26. Elsey, M., & Esedog¯lu, S. (2009). Analog of the total variation denoising model in the context of geometry processing. *Multiscale Modeling & Simulation, 7*(4), 1549–1573.
27. Fisher, P.F.; Pathirana, S. The evaluation of fuzzy membership of land cover classes in the suburban zone. *Remote Sens. Environ.* 1990, *34*, 121–132.
28. Fleming, R. M., Kirkeeide, R. L., Smalling, R. W., Gould, K. L., & Stuart, Y. (1991). Patterns in visual interpretation of coronary arteriograms as detected by quantitative coronary arteriography. *Journal of the American College of Cardiology, 18*(4), 945–951.
29. Flygare, A. A comparison of contextual classification methods using Landsat TM. *Int. J. Remote Sens.* 1997, *18*, 3835–3842.
30. Foody, G.; Cox, D. Sub-pixel land cover composition estimation using a linear mixture model and fuzzy membership functions. *Remote Sens.* 1994, *15*, 619–631.
31. Gharbi, M., Chen, J., Barron, J. T., Hasinoff, S. W., & Durand, F. (2017). Deep bilateral learning for real-time image enhancement. *ACM Transactions on Graphics (TOG), 36*(4), 1–12.
32. Gillespie, A. R., Kahle, A. B., & Walker, R. E. (1987). Color enhancement of highly correlated images. II. Channel ratio and "chromaticity" transformation techniques. *Remote Sensing of Environment, 22*(3), 343–365.
33. Gorelik, N., & Gyftopoulos, S. (2021). Applications of artificial intelligence in musculoskeletal imaging: From the request to the report. *Canadian Association of Radiologists Journal, 72*(1), 45–59.
34. Hamada, Y.; Stow, D.A.; Roberts, D.A.; Franklin, J.; Kyriakidis, P.C. Assessing and monitoring semi-arid shrublands using object-based image analysis and multiple endmember spectral mixture analysis. *Environ. Monit. Assess.* 2013, *185*, 3173–3190.
35. Hansen, K., Baehrens, D., Schroeter, T., Rupp, M., & Müller, K. R. (2011). Visual interpretation of kernel based prediction models. *Molecular Informatics, 30*(9), 817–826.
36. Hay, G. J., Niernann, K. O., & Goodenough, D. G. (1997). Spatial thresholds, image-objects, and upscaling: A multiscale evaluation. *Remote Sensing of Environment, 62*(1), 1–19.
37. Hedblom, M. M., Kutz, O., & Neuhaus, F. (2015). Choosing the right path: Image schema theory as a foundation for concept invention. *Journal of Artificial General Intelligence, 6*(1), 21–54.

38. Helfat, C. E., & Eisenhardt, K. M. (2004). Inter - temporal economies of scope, organizational modularity, and the dynamics of diversification. *Strategic Management Journal, 25*(13), 1217–1232.
39. Honkavaara, E., Arbiol, R., Markelin, L., Martinez, L., Cramer, M., Bovet, S., & Veje, N. (2009). Digital airborne photogrammetry—A new tool for quantitative remote sensing? – A state-of-the-art review on radiometric aspects of digital photogrammetric images. *Remote Sensing, 1*(3), 577–605.
40. Hu, W., Pang, J., Liu, X., Tian, D., Lin, C. W., & Vetro, A. (2021). Graph signal processing for geometric data and beyond Theory and applications. *IEEE Transactions on Multimedia, 24,* 3961–3977.
41. Jeevanantham, D., Dyszuk, E., & Bartlett, D. (2015). The manual ability classification system: A scoping review. *Pediatric Physical Therapy, 27*(3), 236–241.
42. Justice, C.; Townshend, J. A comparison of unsupervised classification procedures on Landsat MSS data for an area of complex surface conditions in Basilicata, Southern Italy. *Remote Sens. Environ.* 1982, *12*, 407–420.
43. Kahn Jr, C. E., Kovatsis, P. G., Messersmith, R. N., & Lehr, J. L. (1989). Automated entry of radiology re□uisition information with artificial intelligence techniques. *American Journal of Roentgenology, 153*(5), 1085–1088.
44. Kallipolitis, A., Revelos, K., & Maglogiannis, I. (2021). Ensembling EfficientNets for the classification and interpretation of histopathology images. *Algorithms, 14*(10), 278.
45. Kashani, A. G., Olsen, M. J., Parrish, C. E., & Wilson, N. (2015). A review of LiDAR radiometric processing: From ad hoc intensity correction to rigorous radiometric calibration. *Sensors, 15*(11), 28099–28128.
46. King, R. L., & Wang, J. (2001, July). A wavelet-based algorithm for pan-sharpening Landsat 7 imagery. In *IGARSS 2001. Scanning the Present and Resolving the Future. Proceedings. IEEE 2001 International Geoscience and Remote Sensing Symposium* (Vol. 2, pp. 849–851).
47. Klemas, V. V. (2009). The role of remote sensing in predicting and determining coastal storm impacts. *Journal of Coastal Research, 25*(6), 1264–1275.
48. Koo, K. M., & Cha, E. Y. (2017). Image recognition performance enhancements using image normalization. *Human-centric Computing and Information Sciences, 7,* 1–11.

49. Kuemmerle, T.; Radeloff, V.C.; Perzanowski, K.; Hostert, P. Cross-border comparison of land cover and landscape pattern in Eastern Europe using a hybrid classification techni□ue. Remote Sens. Environ. 2006, 103, 449–464.

50. Kumar, R.; Nandy, S.; Agarwal, R.; Kushwaha, S.P.S. Forest cover dynamics analysis and prediction modeling using logistic regression model. *Ecol. Indic.* 2014, *45*, 444–455.

51. Kundel, H. L. (1990). Visual cues in the interpretation of medical images. *Journal of Clinical Neurophysiology, 7*(4), 472–483.

52. Kwon, B., Huh, J., Lee, K., & Lee, S. (2020). Optimal camera point selection toward the most preferable view of 3-D human pose. *IEEE Transactions on Systems, Man, and Cybernetics: Systems, 52*(1), 533–553.

53. Lai, V. S., Wong, B. K., & Cheung, W. (2002). Group decision making in a multiple criteria environment: A case using the AHP in software selection. *European Journal of Operational Research, 137*(1), 134–144.

54. Lausch, A., Erasmi, S., King, D. J., Magdon, P., & Heurich, M. (2016). Understanding forest health with remote sensing—Part I: A review of spectral traits, proccsses, and remote-sensing characteristics. *Remote Sensing, 8*(12), 1029.

55. Lee, J. S. (1980). Digital image enhancement and noise filtering by use of local statistics. *IEEE Transactions on Pattern Analysis and Machine Intelligence, (2),* 165–168.

56. Levenson, R. M., & Mansfield, J. R. (2006). Multispectral imaging in biology and medicine: Slices of life. *Cytometry Part A: The Journal of the International Society for Analytical Cytology, 69*(8), 748–758.

57. Li, M., Zang, S., Zhang, B., Li, S., & Wu, C. (2014). A review of remote sensing image classification techni□ues: The role of spatio-contextual information. *European Journal of Remote Sensing, 47*(1), 389–411.

58. Lo, C.; Choi, J. A hybrid approach to urban land use/cover mapping using Landsat 7 Enhanced Thematic Mapper plus (ETM+) images. Int. J. Remote Sens. 2004, 25, 2687–2700.

59. Lunetta, R.S.; Johnson, D.M.; Lyon, J.G.; Crotwell, J. Impacts of imagery temporal frequency on land-cover change detection monitoring. *Remote Sens. Environ.* 2004, *89*, 444–454.

60. Manandhar, R.; Odeh, I.O.; Ancev, T. Improving the accuracy of land use and land cover classification of Landsat data using post-classification enhancement. Remote Sens. 2009, 1, 330–344.

61. Marceau, D. J., Howarth, P. J., & Gratton, D. J. (1994). Remote sensing and the measurement of geographical entities in a forested environment. 1. The scale and spatial aggregation problem. *Remote Sensing of Environment, 49*(2), 93–104.

62. Marcus, W. A., & Fonstad, M. A. (2010). Remote sensing of rivers: The emergence of a subdiscipline in the river sciences. *Earth Surface Processes and Landforms, 35*(15), 1867–1872.

63. Mayes, M.T.; Mustard, J.F.; Melillo, J.M. Forest cover change in Miombo Woodlands: Modeling land cover of African dry tropical forests with linear spectral mixture analysis. *Remote Sens. Environ.* 2015, *165*, 203–215.

64. McNicholas, S., Potterton, E., Wilson, K. S., & Noble, M. E. M. (2011). Presenting your structures: The CCP4mg molecular-graphics software. *Acta Crystallographica Section D: Biological Crystallography, 67*(4), 386–394.

65. Melgani, F.; Al Hashemy, B.A.; Taha, S.M. An explicit fuzzy supervised classification method for multispectral remote sensing images. *IEEE Trans. Geosci. Remote Sens.* 2000, *38*, 287–295.

66. Mellet, E., Tzourio, N., Crivello, F., Joliot, M., Denis, M., & Mazoyer, B. (1996). Functional anatomy of spatial mental imagery generated from verbal instructions. *Journal of Neuroscience, 16*(20), 6504–6512.

67. Milkman, K. L., Chugh, D., & Bazerman, M. H. (2009). How can decision-making be improved? *Perspectives on Psychological Science, 4*(4), 379–383.

68. Mota, G.L.A.; Feitosa, R.Q.; Coutinho, H.L.C.; Liedtke, C.-E.; Müller, S.; Pakzad, K.; Meirelles, M.S.P. Multitemporal fuzzy classification model based on class transition possibilities. ISPRS J. Photogramm. Remote Sens. 2007, 62, 186–200.

69. Newman, M.E.; McLaren, K.P.; Wilson, B.S. Comparing the effects of classification techni□ues on landscape-level assessments: Pixel-based versus object-based classification. *Int. J. Remote Sens.* 2011, *32*, 4055–4073.

70. Olshannikova, E., Ometov, A., Koucheryavy, Y., & Olsson, T. (2015). Visualizing Big Data with augmented and virtual reality: Challenges and research agenda. *Journal of Big Data, 2,* 1–27.

71. Ortiz, S. H. C., Chiu, T., & Fox, M. D. (2012). Ultrasound image enhancement: A review. *Biomedical Signal Processing and Control, 7*(5), 419–428.

72. Passamonti, F., & Maffioli, M. (2016). Update from the latest WHO classification of MPNs: A user's manual. *Hematology 2014, the American Society of Hematology Education Program Book, 2016*(1), 534–542.

73. Pavlovic, V. I., Sharma, R., & Huang, T. S. (1997). Visual interpretation of hand gestures for human-computer interaction: A review. *IEEE Transactions on Pattern Analysis and Machine Intelligence, 19*(7), 677–695.

74. Pelagotti, A., Del Mastio, A., De Rosa, A., & Piva, A. (2008). Multispectral imaging of paintings. *IEEE Signal Processing Magazine, 25*(4), 27–36.

75. Peterson, S.H.; Stow, D.A. Using multiple image endmember spectral mixture analysis to study chaparral regrowth in southern California. *Int. J. Remote Sens.* 2003, *24*, 4481–4504.

76. Pottmann, H., Wallner, J., Huang, Q. X., & Yang, Y. L. (2009). Integral invariants for robust geometry processing. *Computer Aided Geometric Design, 26*(1), 37–50.

77. Poursanidis, D.; Chrysoulakis, N.; Mitraka, Z. Landsat 8 vs. Landsat 5: A comparison based on urban and peri-urban land cover mapping. *Int. J. Appl. Earth Obs. Geoinf.* 2015, *35*(Part B), 259–269.

78. Powell, R.L.; Roberts, D.A.; Dennison, P.E.; Hess, L.L. Sub-pixel mapping of urban land cover using multiple endmember spectral mixture analysis: Manaus, Brazil. *Remote Sens. Environ.* 2007, *106*, 253–267.

79. Pullanikkatil, D.; Palamuleni, L.; Ruhiiga, T. Assessment of land use change in Likangala River catchment, Malawi: A remote sensing and DPSIR approach. *Appl. Geogr.* 2016, *71*, 9–23

80. Qi, Y., & Wu, J. (1996). Effects of changing spatial resolution on the results of landscape pattern analysis using spatial autocorrelation indices. *Landscape Ecology, 11,* 39–49.

81. Rahman, Z. U., Jobson, D. J., & Woodell, G. A. (2004). Retinex processing for automatic image enhancement. *Journal of Electronic Imaging, 13*(1), 100–110.

82. Raup, B., Kääb, A., Kargel, J. S., Bishop, M. P., Hamilton, G., Lee, E., & Helm, C. (2007). Remote sensing and GIS technology in the Global Land Ice Measurements from Space (GLIMS) project. *Computers & Geosciences, 33*(1), 104–125.

83. Raziei, T., Bordi, I., & Pereira, L. S. (2013). Regional drought modes in Iran using the SPI: The effect of time scale and spatial resolution. *Water Resources Management, 27,* 1661–1674.

84. Roberts, D.A.; Gardner, M.; Church, R.; Ustin, S.; Scheer, G.; Green, R.O. Mapping chaparral in the Santa Monica Mountains using multiple

endmember spectral mixture models. *Remote Sens. Environ.* 1998, *65*, 267–279.

85. Rosenbaum, D. A. (2021). The ultimate tool: The body, planning of physical actions, and the role of mental imagery in choosing motor acts. *Topics in Cognitive Science, 13*(4), 777–799.

86. Saaty, T. L. (1990). How to make a decision: The analytic hierarchy process. *European Journal of Operational Research, 48*(1), 9–26.

87. Samal, D.R.; Gedam, S.S. Monitoring land use changes associated with urbanization: An object based image analysis approach. *Eur. J. Remote Sens.* 2015, *48*, 85–99.

88. Santos, A. B., de Albuquerque Araújo, A., & Menotti, D. (2013). Combining multiple classification methods for hyperspectral data interpretation. *IEEE Journal of Selected Topics in Applied Earth Observations and Remote Sensing, 6*(3), 1450–1459.

89. Schaaf, A. N., Dennison, P. E., Fryer, G. K., Roth, K. L., & Roberts, D. A. (2011). Mapping plant functional types at multiple spatial resolutions using imaging spectrometer data. *GIScience & Remote Sensing, 48*(3), 324–344.

90. Schuster, D. M., & Gale, M. E. (1996). The malady of incomplete, inadequate, and inaccurate radiology requisition histories: A computerized treatment. *AJR. American Journal of Roentgenology, 167*(4), 855–859.

91. Schwarz, G. F., Rybach, L., & Klingele, E. E. (1994). Data processing and mapping in airborne radiometric surveys. *ITC Journal, (1),* 34–39.

92. Shabala, S. (2003). Screening plants for environmental fitness: Chlorophyll fluorescence as a "holy grail" for plant breeders. In *Advances in Plant Physiology* (Vol. 5, pp. 287–340). Jodhpur: Scientific Publishers, India.

93. Shogenji, R., Kitamura, Y., Yamada, K., Miyatake, S., & Tanida, J. (2004). Multispectral imaging using compact compound optics. *Optics Express, 12*(8), 1643–1655.

94. Sikder, I.U. Knowledge-based spatial decision support systems: An assessment of environmental adaptability of crops. *Expert Syst. Appl.* 2009, *36*, 5341–5347.

95. Sloan, S. Historical tropical successional forest cover mapped with Landsat MSS imagery. *Int. J. Remote Sens.* 2012, *33*, 7902–7935.

96. Steiniger, S., & Hunter, A. J. (2013). The 2012 free and open source GIS software map—A guide to facilitate research, development, and adoption. *Computers, Environment and Urban Systems, 39,* 136–150.

97. Stuckens, J.; Coppin, P.R.; Bauer, M.E. Integrating contextual information with per-pixel classification for improved land cover classification. *Remote Sens. Environ.* 2000, *71*, 282–296.

98. Svatonova, H. (2016). Analysis of visual interpretation of satellite data. *The International Archives of the Photogrammetry, Remote Sensing and Spatial Information Sciences, 41,* 675–681.

99. Swain, P.H.; Vardeman, S.B.; Tilton, J.C. Contextual classification of multispectral image data. *Pattern Recognit.* 1981, *13*, 429–441.

100. Tailor, A.; Cross, A.; Hogg, D.C.; Mason, D.C. Knowledge-based interpretation of remotely sensed images. *Image Vis. Comput.* 1986, *4*, 67–83.

101. Tasdizen, T., Whitaker, R., Burchard, P., & Osher, S. (2003). Geometric surface processing via normal maps. *ACM Transactions on Graphics (TOG), 22*(4), 1012–1033.

102. Taudes, A. (1998). Software growth options. *Journal of Management Information Systems, 15*(1), 165–185.

103. Théau, J.; Peddle, D.R.; Duguay, C.R. Mapping lichen in a caribou habitat of Northern Quebec, Canada, using an enhancement_classification method and spectral mixture analysis. *Remote Sens. Environ.* 2005, *94*, 232–243.

104. Toll, D.L. Effect of Landsat Thematic Mapper sensor parameters on land cover classification. *Remote Sens. Environ.* 1985, *17*, 129–140.

105. Ton, J.; Sticklen, J.; Jain, A.K. Knowledge-based segmentation of Landsat images. IEEE Trans. Geosci. Remote Sens. 1991, 29, 222–232.

106. Toutin, T. (2004). Geometric processing of remote sensing images: Models, algorithms, and methods. *International Journal of Remote Sensing, 25*(10), 1893–1924.

107. Treitz, P., & Howarth, P. (2000). High spatial resolution remote sensing data for forest ecosystem classification: An examination of spatial scale. *Remote Sensing of Environment, 72*(3), 268–279.

108. Vailaya, A., Jain, A., & Zhang, H. J. (1998). On image classification: City images vs. landscapes. *Pattern Recognition, 31*(12), 1921–1935.

109. Vieira, I.C.G.; de Almeida, A.S.; Davidson, E.A.; Stone, T.A.; Reis de Carvalho, C.J.; Guerrero, J.B. Classifying successional forests using Landsat spectral properties and ecological characteristics in Eastern Amazônia. *Remote Sens. Environ.* 2003, *87*, 470–481.

110. Vijayaraj, V., Younan, N. H., & O'Hara, C. G. (2006). Quantitative analysis of pan-sharpened images. *Optical Engineering, 45*(4), 046202–046202.

111. Wang, L. Fuzzy supervised classification of remote sensing images. IEEE Trans. Geosci. Remote Sens. 1990, 28, 194–201.
112. Wang, L.; Newkirk, R. A knowledge-based system for highway network extraction. *IEEE Trans. Geosci. Remote Sens.* 1988, *26*, 525–531.
113. Wang, P. S., Fu, X. M., Liu, Y., Tong, X., Liu, S. L., & Guo, B. (2015). Rolling guidance normal filter for geometric processing. *ACM Transactions on Graphics (TOG), 34*(6), 1–9.
114. Wang, Y. Q. (2014). MeteoInfo: GIS software for meteorological data visualization and analysis. *Meteorological Applications, 21*(2), 360–368.
115. White, C. W., Wright, C. B., Doty, D. B., Hiratza, L. F., Eastham, C. L., Harrison, D. G., & Marcus, M. L. (1984). Does visual interpretation of the coronary arteriogram predict the physiologic importance of coronary stenosis? *New England Journal of Medicine, 310*(13), 819–824.

Chapter 6

ESTIMATION OF CROP WATER PRODUCTIVITY VIA REMOTE SENSING

Contents

6.1. INTRODUCTION

One of the major difficulties facing the globe today is feeding everyone as the population grows. A UN estimate from DESA, or the Economic and Social Affairs Department, states that by 2050, the Earth will house 9.3 billion people on the planet, suggesting that production of food must rise to roughly 70–100%. It is important to acknowledge that the substantial increase in food demand will necessitate additional land and arable water resources (Kawabata et al., 2020). Drought and crop water stress are the main causes of decreased food yields, particularly in locations that are arid or semi-arid (Wei et al., 2021). Agriculture uses over 70% of the freshwater resources on Earth. Global agricultural water use is thought to touch 3200 billion cubic meters by 2050 owing to population growth. A shortage of water resources is predicted to affect 67% of the world's population because of the enormous agricultural water demand. Crop water productivity or CWP is a crucial metric for characterizing the interplay between water resources and food. It is expressed as the ratio of crop output to water consumption. Thus, for global food security, strengthening the CWP and achieving a balance between the production of food and water resources is essential.

The multidisciplinary convergence of hydrology and agriculture is involved in the estimate of CWP. Remote sensing has found significant use in hydrology and agriculture since the launch of the first weather satellite in the 1960s by the US successfully in the 1960s. RS provides a more economical approach to CWP estimation for wider physical regions with higher spatial and temporal resolutions as compared to in-situ approaches. Currently, the most popular remote sensing platforms are unmanned aerial vehicles (UAV), manned aerial vehicles (MAV), and satellites along with ground instrumentation. Based on these remote sensing platforms, numerous algorithms with a range of application scenarios have been created for the estimate of agricultural and hydrological parameters. These techniques include the following: the rectangular hyperbola model for crop net primary productivity (NPP) estimation; the light use efficiency (LUE) model in estimation of crop gross primary productivity or GPP; the machine learning-based model for regional soil moisture content monitoring, and the surface energy balance algorithm of land (SEBAL) for regional evapotranspiration estimation (Ahmed et al., 2010).

For CWP estimation, a comprehensive examination of the RS methods as they relate to hydrology and agriculture is required. Unfortunately, there is currently no benchmark available for the CWP estimation method. This section reviews the definition, methods, applicability situations, and accuracy evaluation of the RS approach for CWP estimate (Chiraz et al., 2022).

6.2. CROP WATER PRODUCTIVITY DEFINITION

6.2.1. Crop Water Productivity

The utilization of water in agriculture can be broadly classified into the following processes: 1) irrigation, 2) water transmission, and 3) crop water use. Water is efficiently transported from the source to the field using hydraulic engineering structures, and during irrigation, crops are supplied with this water. The water is absorbed by the crops and eventually converted into biomass, contributing to crop yield as part of the crop water use process. The canal water use efficiency (CWUE), crop water productivity (CWP), and field water use efficiency (FWUE) are used to describe these three processes. Users can easily quantify CWP using the measurable properties of these three indicators. FWUE and CWUE are commonly employed to assess the efficiency of irrigation technology. Equation (1) defines CWP as the crop yield produced per unit of crop water consumed:

$$CWP(kg/m^3) = \frac{CY(kg/ha)}{10 \times \sum ET(mm)} \tag{1}$$

where; *CY* represents crop yield; and $\sum$ET denotes the cumulative water consumed by the crop throughout the growth (Kang et al., 2009).

In contrast to FWUE and CWUE, CWP takes into account every possible source of water during crop growth, involving groundwater and precipitation in addition to irrigation. Thus, CWP applies to both irrigated and rainfed farmland, as mentioned in the study by Blatchford et al. (2019). Throughout recent years, CWP has gained popularity as a reliable measure for assessing agricultural efficacy.

The UN Sustainable Development Goals (SDGs) utilized CWP to define the objective of developing agricultural water management (Ban, 2016). When making decisions on food production and field management, CWP can set a standard or specify an objective. Additionally, it is being employed to evaluate the agricultural water usage efficiency policies in the long run. Thus, CWP takes center stage in this chapter.

6.2.2. Crop Yield

Understanding the factors that influence crop yield is crucial for successful crop management. Three key variables that are closely linked to crop yield are 1). Aboveground biomass (AGB); 2). Net primary productivity (NPP); and 3). Gross primary productivity (GPP). The quantitative relationship between them is provided by Equations 2–5 and Figure 6.1. The GPP ($gC{\cdot}m^{-2}{\cdot}day^{-1}$) represents the total organic carbon quantity resolved by the plants via the process of photosynthesis. There are three types of present remote sensing-based GPP estimate methods: models dependent on LUE, and models based on processes along empirical models (Lobell et al., 2009).

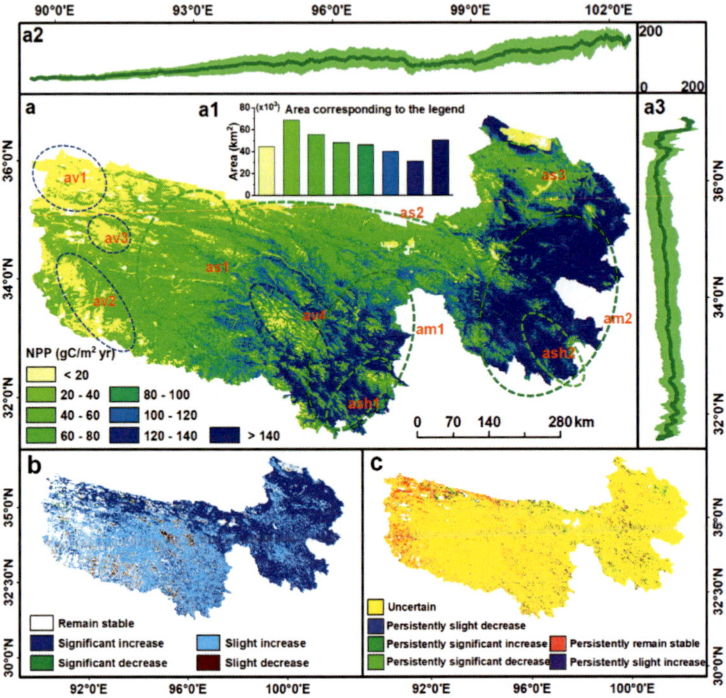

Figure 6.1. *Distribution of the net primary productivity (NPP) and its changing trend in the Three-River Headwaters region, including (a) the average NPP from 2001 to 2020 [av: alpine vegetation; as: alpine steppe; am: alpine meadow; ash: alpine shru□land]; (a1) the area percentage corresponding to the eight classes in (a); (a2) the average NPP and its one standard deviation (SD) along the longitude; (a3) the average NPP and its SD along the latitude; (□) the changing trend of NPP from 2001 to 2020; (c) the changing characteristics of the NPP from 2001 to 2020.*

Source: Liu, W., Yuan, Y., Li, Y., Li, R., & Jiang, Y. (2023). Net primary productivity estimation using a modified MOD17A3 model in the Three-River Headwaters region. Agronomy, 13(2), 431. https://doi.org/10.3390/agronomy13020431 (Creative Commons License).

Statistical methods are used to estimate GPP by fitting relationships between RS data and ground carbon flux measurements, such as chlorophyll fluorescence and VIs. Process-based models, such as the BESS or Breathing Earth System Simulator model and the PML or Penman-Monteith-Leuning model, estimate GPP from the perspective of physiological and ecological processes (Condon et al., 2002). Based on the LUE model, GPP is found by multiplying photosynthetic efficiency by photosynthetically active radiation or PAR that the plant absorbs:

$$GPP = PAR \times f_{PAR} \times \varepsilon$$

In this equation,   denotes the photosynthetic efficiency; f_{PAR} is the fraction of *PAR* given by $MJ \cdot m^{-2} \cdot day^{-1}$ absorbed by the canopy. A lot of people use LUE-based models these days. Some of the most common ones are the Vegetation Production Model or the VPM, the Vegetation Production and Respiration Model or the VPRM, and the Eddy Covariance-Light Use Efficiency or EC-LUE. Others are the two-stage LUE or the TS-LUE model and the Moderate Resolution Imaging Spectroradiometer-GPP, known as MOD17. Different LUE-based algorithms use different parameterization schemes, and they are good for estimating the GPP in various places (Cabas et al., 2010).

The net primary productivity (NPP), expressed in $gC \cdot m^{-2} \cdot day^{-1}$, represents the total amount of fixed organic carbon remaining after deducting the autotrophic respiration (AR) of plants, also measured in $gC \cdot m^{-2} \cdot day^{-1}$:

$$NPP = GPP - AR$$

The conversion of NPP to AGB ($kg \cdot ha^{-1}$) was often done using the following empirical formula:

$$AGB = k \times \sum NPP$$

Where; *k* represents the aboveground fraction; and $\sum NPP$ represents the cumulative *NPP* of the crop's entire growth period. According to Walter et al. (2018), the crop yield (CY, $kg \cdot ha^{-1}$) may be calculated as:

$$CY = \frac{HI \times AGB}{1 - \theta_g}$$

The harvest index, which comes to be the conversion ratio of conversion of grain yield from the AGB, is determined by the crop species and growth environment and species of crops. The purpose of θg, or standard grain moisture, is to standardize yields to identical water content standards (Challinor et al., 2014).

6.2.3. Evapotranspiration

Crop evapotranspiration (ET_a) is a key factor in determining crop water consumption. It encompasses both plant transpiration (T) and soil evaporation (E). Nevertheless, remote sensing-based methods often estimate these two processes together due to the challenge of separating E and T, as noted by Blatchford et al. (2019). While various empirical formulas are applied to estimate E and T using the P-M model and produced MOD16 ET products, there is still room for enhancing the accuracy of E or T, respectively (Hargreaves & Samani, 1982).

It is worth mentioning that certain studies have used two related evapotranspiration theories, namely potential evapotranspiration (ET_p) and reference crop evapotranspiration (ET_o), interchangeably. Nevertheless, Xiang et al. (2020) propose alternative definitions for these terms. The ETo value represents the estimated evapotranspiration of a reference crop with specific characteristics, such as albedo and height of 0.23 and 0.12 m respectively, followed by the surface resistance of about 70 $s{\cdot}m^{-1}$. It comes out to be the grass ET_a that grows uniformly, quickly, and tightly without being affected by pests, illnesses, or stress. ET_p, on the other hand, is the actual demand for atmospheric evaporation, also called atmospheric evaporation capacity, or the highest ET_a value when conditions are ideal. ET_p was first utilized to assess the extent of drought. Previous research has provided a range of equations for estimating ET_o and ET_p. You can find more information in the study conducted by Xiang et al. (2020).

6.3. CROP WATER PRODUCTIVITY (CWP) ESTIMATION FRAMEWORK

As explained in Equation 1, the CWP is calculated based on the values of CY and ET_a. The remote sensing (RS) techniques for CY and ET_a estimates are briefly explained in this section.

6.3.1. The Crop Yield (CY) Estimation

A pretty accurate CY estimation is essential for determining CWP. In addition, accurate estimation of crop yield is crucial for agrarian production. Utilizing RS technology allows for an economic method to predict crop yield on a larger scale and then map its distribution. There are two main types of methods used to estimate CY: (i) statistical regression (SR) techniques along (ii) crop model data assimilation (DA) techniques (Bouman, 1995).

6.3.1.1. Statistical Regression or SR Methods

The statistical regression methods obtain an approximate of crop yields by matching the actual yield and the parameters that were sensed from a distance. Some basic VIs, such as the NDVI, initially were employed for crop yield estimation using a straightforward linear function. According to these methodologies, VIs could be used to indicate the status of crop growth and serve as the foundation for yield estimation. In addition to vegetation indices, other factors such as crop phenology; information on crop structure; meteorological data; and temperature of the surface were considered when estimating crop yield.

In addition, before estimating crop yield, certain studies have transformed vegetation indices and meteorological data into intermediate indicators such as ET_a and GPP. The physiological state of the plant is described by these intermediate markers, which have a direct impact on CY. Yuan et al. (2016) created an EC-LUE model for estimating the world GPP and thus, CY. In a study conducted by Sibley et al. (2014), the researchers utilized absorbed PAR as an important metric and mixed it with LUE to make estimations on maize yield in Nebraska. Moreover, extensive research has discovered a robust association between CY and the following variables: leaf area index or the LAI, cumulative temperature, crop evaporation stress, GPP, and ET_a. As is often the case, the models provide estimates for the crop yields that may contain errors stemming from the parameters and variable uncertainty. However, incorporating the intermediate indicators within the statistical regression models has the potential to enhance the estimation of crop yields.

Machine learning or ML is also utilized in statistical regression methods to estimate crop yields, as demonstrated by various studies. The connection between data obtained remotely and yield is often non-linear, as highlighted in studies by Weiss et al. (2019). To get an accurate analysis of this relationship, advanced machine-learning algorithms like random forest and neural networks have proven to be effective. In addition, choosing the right vegetation indices during specific stages of plant growth is crucial for accurately estimating crop yield. In most cases, the final condition of the crop would give a rather precise representation of the crop yields since it has already been determined. Earlier prediction of yield is highly valuable for making agricultural management decisions. Hence, the timing of predictions involves finding a balance between its accuracy and practicality. As demonstrated, the stage of seed-filling was found to be the most optimal to estimate the yield of soybeans. In a study, it was discovered that the prediction of winter wheat yield, done around 2–4 weeks prior to harvest, showed promising results. Ideally, the prediction time should be aligned with the specific phenological period of the crop. In addition, there are variations in the regional scales of the crop's phenological period. Although the importance of phenological information as a gridded dataset in yield prediction cannot be overstated, only limited research has considered this data on a regional scale (McIntyre et al., 1966).

In conclusion, the use of SR methods offers a straightforward, affordable, and accurate approach to estimating CY, without the need for extensive parameters and calibration. The general flowchart for SR approaches is displayed in Figure 6.2. It is important to highlight that the machine learning models have been specifically designed to handle spatial heterogeneity and variation caused by factors such as planting patterns, including fertilization. Planting density, irrigation schedules, and crop variety. Thus, delving into the parameters of spatial heterogeneity is crucial for addressing this problem (Hosseini et al., 2017).

6.3.1.2. Methods Based on Data Assimilation (DA)

Several crop growth models, such as WOFOST, DAISY, DSSAT, and AquaCrop have been thoroughly examined by Jin et al. (2018). These models are widely employed to estimate crop yield. These models were initially designed to simulate crop growth at a small scale, but they have been increasingly used to estimate yield across larger areas when combined with data obtained through remote sensing. Using RS information, the Data assimilation-based techniques incorporate the state variables of the crop model, such as LAI or canopy coverage, to provide a detailed description of the crop growth status at a regional scale (Arcucci et al., 2021). A typical flowchart for DA-based Crop yield estimate techniques is displayed in Figure 6.3.

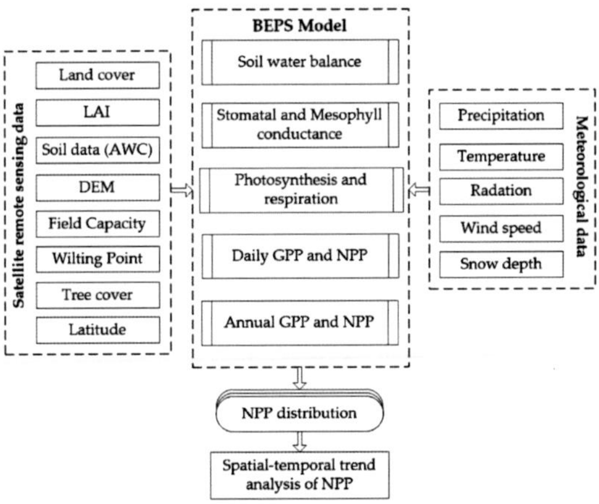

Figure 6.2. *The framework of the boreal ecosystem productivity simulator (BEPS) model. [LAI: leaf area index; AWC: availa□le soil water-holding capacity; DEM: digital elevation model; GPP: gross primary productivity; NPP: net primary productivity].*

Source: Nanzad, L., Zhang, J., Batdelger, G., Pangali Sharma, T. P., Koju, U. A., Wang, J., & Nabil, M. (2021). Analyzing NPP response of different rangeland types to climatic parameters over Mongolia. Agronomy, 11(4), 647. https://doi.org/10.3390/agronomy11040647 (Creative Commons License).

The crop model lays the foundation of the data assimilation methods. A crop model is frequently used to explore the relationship between the growth environment and the status of the crops, as well as in the simulation of crop growth, which is usually defined using numerous indicators throughout the entire growth stage.

The first of these models was put forward in the 1960s. A multitude of crop models have been created over the last 60 years, including APSIM, DSSAT, MONICA, and AquaCrop. Irrespective of the chosen model, acquiring the parameters is a critical process. Acquiring parameters from various regions poses a significant challenge for crop model performance, particularly for parameters that exhibit spatial heterogeneity, such as varieties of crops, amount of irrigation, and moisture of soil. The specific parameters of several widely used crop models are displayed in Table 6.1. Every crop model has its own set of advantages and disadvantages, as the means of assimilation and requirements of parameters vary. Therefore, it is important to choose the most suitable model based on the specific application scenario (Zhang et al., 2011).

By combining dynamic models (such as crop growth models) with data (such as remotely sensed indicators), the DA-based approaches estimate CY. Remote sensing observations have a direct impact on the accuracy of yield estimation.

The DA-based techniques have been divided into three groups: updating, forcing, and calibration methods (Jin et al., 2018). To achieve the best consistency between simulated indicators or state variables and remotely sensed indicators, calibration methods modify the starting parameters of crop models (Figure 6.4). By implementing these methods, the accuracy of RS data can be significantly improved, minimizing error accumulation and propagation. Some common ways to calibrate are the Very Fast Annealing Algorithm; the Particle Swarm Optimization Algorithm; the Least-Squares Method; and the Maximum Likelihood Solution.

Forcing methods, which are straightforward and do not require complex assimilation algorithms, use remotely detected indicators to directly substitute generated indications (Figure 6.4). In addition, the use of RS indicators instead of model-simulated indicators necessitates a higher level of accuracy for the RS indicators (Wikle & Berliner, 2007). Crop model simulation indicators are regularly updated through the use of remotely sensed indicators in updating methods (Figure 6.4). Two common updating techniques are Four-Dimensional Variational Data Assimilation or 4DVAR) and the Ensemble Kalman Filter or the EnKF (Carrassi et al., 2017).

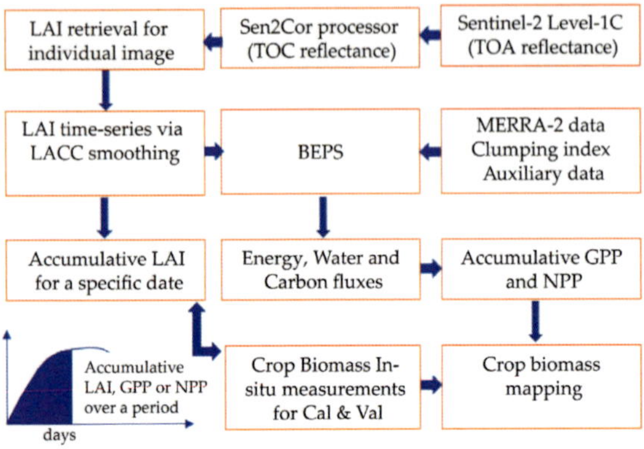

Figure 6.3. *A diagram of the data processing flow for mapping crop biomass using sentinel-2 data and the boreal ecosystems productivity simulator (BEPS).*

Source: He, L., Wang, R., Mostovoy, G., Liu, J., Chen, J. M., Shang, J., Liu, J., McNairn, H., & Powers, J. (2021). Crop biomass mapping based on ecosystem modeling at a regional scale using high-resolution Sentinel-2 data. Remote Sensing, 13(4), 806. https://doi.org/10.3390/rs13040806 (Creative Commons License).

6.3.1.3. Ground Measurements

The connection between GPP and CY is illustrated in Section 6.2. Therefore, the yield estimations may be done through the measurement of carbon fluxes obtained from in-situ field measurements. The eddy covariance (EC) and chamber methods are widely used for this purpose. According to Baldocchi (2003), the Eddy Covariance method usually covers just a few hectares for measuring carbon fluxes. The chamber method is limited to measuring the concentration of gas within the chamber-enclosed area. Therefore, the practicality of this method in measuring carbon fluxes in the field is somewhat restricted. The EC approach uses the covariance between vertical wind velocity fluctuations and CO_2 to determine the exchange rate through the interface of the atmospheric and crop canopy interface. The two approaches performed similarly in terms of measurement accuracy. Nevertheless, only a small number of research used such ground measurements for crop yields, and even fewer examined how well the EC approaches performed in comparison to the real crop yields. Such techniques may only be feasible theoretically. In addition, it is important to mention that the cumulative GPP during the entire crop growth is utilized in estimating the Crop Yield (CY), as demonstrated in Equations 2–5. However, it is worth noting that the observation area for EC measurements may change depending on the weather conditions. Therefore, acquiring the cumulative GPP in a particular region can be quite challenging (Balzarolo et al., 2011).

Table 6.1. *Examples of Remote Sensing Information, Derived from Sentinel-2, Used for Crop Monitoring and Management, and the Methodological Approach Used*

S2 for Agriculture		**Sensed Trait**	**Method**	**References**
Plant stress	Biotic	Cotton rot root (*Phymatotrichopsis omnivore*)	Classification of affected areas with trained algorithms using bands 2, 3, 4, and 8. S2 imagery calibrated with a UAV. NDVI decreases with increasing infestation, ground, airborne, and satellite multiplatform. Field spectral data and S2-derived Vis. Proposed index: Red edge disease stress index (REDSI), bands 4, 5, and 7.	[Chemura, A.;– Pla, M.; et.al]
		Rice crops and Western Swamphen (*Porphyrio porphyrio*)		
		Hessian fly (*Mayetiola destructor*) infestation in Wheat		
		Coffee leaf rust (*Hemileia vastarix*)		
		Wheat yellow rust (*Puccinia striiformis*)		
	Abiotic	Metal stress in rice	Red-edge S2 bands. S2 VI and OPTRAM soil moisture monitoring.	[Zhang, Z.;– Sun, N.; Mananze, S.; West, H.; et.al]
		Drought		
		Salinity	S2 visible bands, blue and red, are sensitive to soil salinity.	[Davis, E.; et.al]

Management scale	Fertilization	N in crops	Biophysical retrieval of canopy chlorophyll content, also with VIs within the red edge. Assessment of the nutritional status, NNI	[Clevers, J.G.P.W.;–Söderström, M.; et.al]
	Water	Irrigation and hydric requirements (cotton, tomato, wheat, and maize).	S2 vegetation parameters, surface albedo, and crop height for FAO-56 Penman-Monteith ET estimation; red and red-edge bands to predict crop coefficients (Kc). Irrigated and rain-fed cropland differentiation. Combination of S2 and Aquacrop.	[Vanino, S.; Rozenstein, O.; et.al]
	Fields monitoring	Cropland assessments	VIs for cropping practices assessment; regional and nation-scale cropland and crop type classification with S2, time series, and retrieval of biophysical and vegetation radiometric indexes (sen-2Agri).	[Defourny, P.; Lambert, M.J.; Vuolo, F.; Noi, P.T.; et.al]
	Soils	Soil features	Determining soil OM with VIs, and S2 bands. The wavelength of the OM spectral feature in the visible is close to the S2 red band. Classification of soil degradation.	[Castaldi, F.; Žížala, D.; et.al]
	Yield prediction	Empirical models Radiative transfer models (RTM)	VIs and yields, together with climatological data to build a dataset. Fitting techniques (regressions, random forest, machine learning) to predict yields. Coupling with crop functioning models, FAPAR, LAI, SLA, and light use efficiency.	[Gómez, D.; He, L.; et.al]

Source: Segarra, J., Buchaillot, M. L., Araus, J. L., & Kefauver, S. C. (2020). Remote sensing for precision agriculture: Sentinel-2 improved features and applications. Agronomy, 10(5), 641. https://doi.org/10.3390/agronomy10050641.

6.3.2. Evapotranspiration

Earlier investigations have thoroughly reviewed and summarized remote sensing methods for ETa estimations. There are four main categories of methods based on remote sensing for estimating ETa: Ground measurements; statistical methods; Semi-Empirical Formula (SEF); and Surface Energy Balance Residual (SEBR) (Pereira et al., 1999).

6.3.2.1. The Surface Energy Balance Residual or SEBR Method

It works out residual surface energy to determine the latent heat flux (LE) given by W·m^{-2}, and this is the energy utilized in evaporating water from the surface:

$$LE = Rn - G - H$$

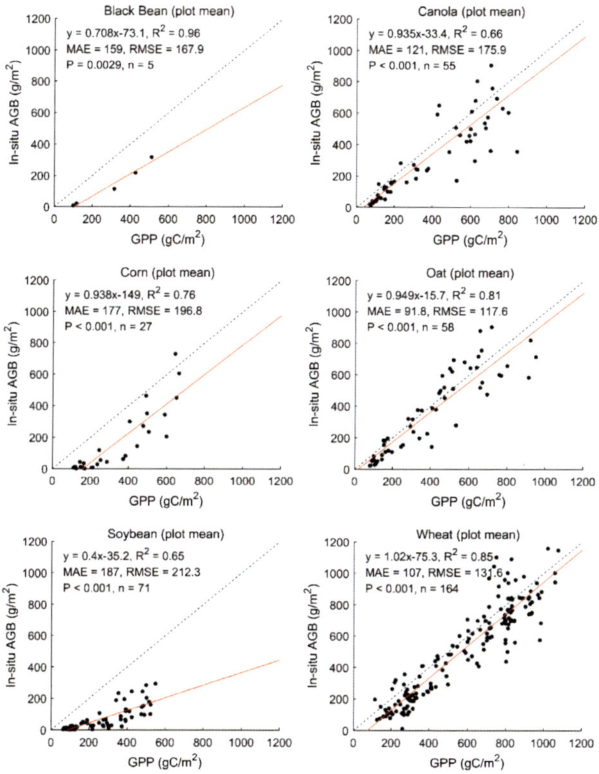

Figure 6.4. *The relationship between BEPS-simulated crop NPP and ground-measured above-ground biomass for each crop type for SMAPVEX16-MB.*

Source: He, L., Wang, R., Mostovoy, G., Liu, J., Chen, J. M., Shang, J., Liu, J., McNairn, H., & Powers, J. (2021). Crop biomass mapping based on ecosystem modeling at a regional scale using high-resolution Sentinel-2 data. Remote Sensing, 13(4), 806. https://doi.org/10.3390/rs13040806.

Net radiation or R_n given by $W{\cdot}m^{-2}$ could be determined using Equation 7, while ground heat flux (G) given by G ($W{\cdot}m^{-2}$) is typically determined via an empirical formula that gives value related to Vegetation Indexes and R_n. E□uation 9 can be used to compute the sensible heat flux, H given by $W{\cdot}m^{-2}$, shown in Equation 8.

$$Rn = (1-a)\frac{Gsc \times \cos\theta \times \tau_{sw}}{d_r^2} - \varepsilon\sigma Ts^4 + \varepsilon_a\sigma Ta^4 - (1-\varepsilon)\varepsilon_a\sigma Ta^4$$

$$G = R_n \times \frac{T_s - 273.16}{a}(0.0032 \times \frac{a}{c} + 0.0032(\frac{a}{c})^2) \times (1 - 0.978NDVI^4)$$

$$H = \frac{\rho_{air}Cp(To - Ta)}{r_a}$$

The table below explains the variables in the equations.

Symbol	Representation
α	Surface albedo
θ	Solar zenith angle
Gsc	Solar constant (1376 $W{\cdot}m^{-2}$)
τsw	Atmospheric transmittance
εa	Atmospheric emissivity
ε	Surface emissivity
σ	Stefan-Boltzmann constant (5.67×10^{-8} $W{\cdot}m^{-2}K^{-4}$)
Ts, Ta, and To	Surface, air, and surface aerodynamic air temperature.
ρair	Air density ($kg{\cdot}m^{-3}$)
Cp	Air-specific heat at constant pressure ($J{\cdot}kg^{-1}{\cdot}K^{-1}$)
ra	Aerodynamic resistance ($s{\cdot}m^{-1}$)

In local time (LT), when transmission occurs before 12:00, c has a value of 0.9; when it occurs between 12:00 and 14:00, the value becomes 1.0; and when it occurs between 14:00 and 16:00, it is 1.1.

The SEBR can be categorized into one- and two-source models based on the soil, whether is distinguished from the canopy in the energy balance system (Prime et al., 2004).

According to Cheng et al. (2012), one-source models consider vegetation and soil together and suppose equal dynamic resistance for them. Initially put forth in the mid-1980s, one-source models substituted remotely sensed T_s for T_o when predicting H (Equation 9) at some specific viewing angle. Two commonly used models in the field are the Surface Energy Balance System or SEBS and the Surface Energy Balance Algorithm of Land or SEBAL. These models are built on a strong theoretical foundation and are highly portable. They have been extensively utilized in estimating ETa by various researchers. It is important to understand that there is a distinction between T_o and T_s obtained remotely. This happens since such T_s generally refers to the radiative temperature of the surface of land on a roughness layer. According to Equation 9, heat and water turbulence in the constant flow layer causes the temperature to close to the surface, as indicated by T_o. In the context of a roughness layer, the exchange of water and heat is primarily influenced by molecular diffusion rather than turbulent exchange. This means that significant gradients are needed for the diffusion process to occur, as observed in the constant flux layer. According to several studies, T_s may be significantly greater during the day than T_o, particularly in areas with little to no vegetation and exposed soil. Therefore, there will be a significant overestimation of H when T_s is used instead of T_o, resulting in an underestimation of LE (Li et al., 2018).

In two-source models, there is a separation of the surfaces into the following parts: plant and bare soils. According to Wang & Dickinson (2012), Ts is a combination of the soil temperature (Tso) and canopy temperature (Tca). Traditional research proposed two-source models for sparse vegetation surfaces. Some examples of two-source models involve the Two Sources Energy Balance (TSEB), Two-source Trapezoid Model for Evapotranspiration (TTME), and Trapezoid Framework-based Evapotranspiration Model (HTEM) and Hybrid Dual-source Scheme. Another approach, developed by Tang and Li, takes into account the different soil water content responses at the upper layer of the surface and the deeper layer of the root zone. A more comprehensive theoretical framework is provided by the Two-source models compared to one-source models. There have been studies that demonstrate the superior accuracy of two-source models compared to one-source models.

Remotely sensed data is usually captured in an instant. Therefore, it is crucial to scale up from an instantaneous to a daily scale. Currently, it is widely acknowledged that the ratios of the ground energy budget fluxes stay relatively constant throughout the day. This allows for the calculation of the ratio based on the real-time satellite transmission data. Additionally, this flux may be calculated on a daily scale, as demonstrated by Crago (1996). According to Jackson et al. (1983), during daylight, the flux ratio given by $LE \cdot S^{-1}$, where S is the shortwave downward radiation) remains almost unchanged. According to Shuttleworth (1989), the constant evaporation fraction theory calculated by $EF = LE(LE + H)^{-1}$ suggests that the EF remains relatively stable throughout the day. This hypothesis has been extensively utilized in the research involving

LE upscaling. In addition, according to studies various methods have been employed for up-scaling LE, including the radiation ratio method; referenced evaporation fraction method; constant decoupling factor method; and cosine curve method. The physical constant decoupling factor method, given by Tang & Li (2017), effectively addresses the disparity of instantaneous EF during the satellite overpass times and during other timeframes. This method has been proven to be reliable and outperforms the constant evaporation fraction approach (Figure 6.5).

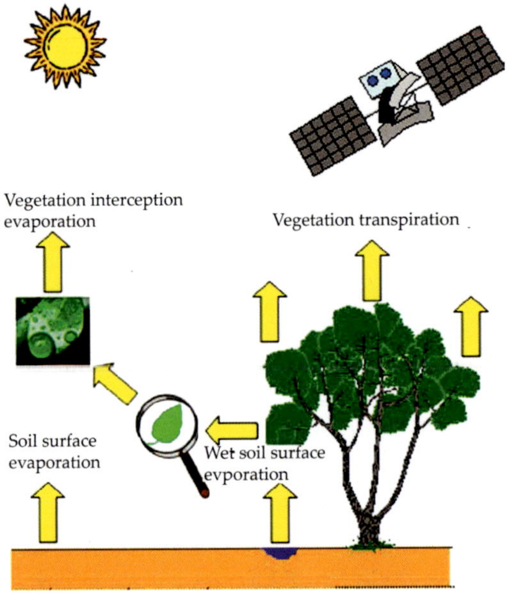

Figure 6.5. *Schematic diagram of the modified satellite-□ased Priestley-Taylor model.*

Source: Yao, Y., Liang, S., Zhao, S., Zhang, Y., Qin, Q., Cheng, J., Jia, K., Xie, X., Zhang, N., & Liu, M. (2014). Validation and application of the modified satellite-based Priestley-Taylor algorithm for mapping terrestrial evapotranspiration. Remote Sensing, 6(1), 880–904. https://doi.org/10.3390/rs6010880.

6.3.2.2. Measurements on Ground

Ground measurement instruments, like the EC methods and scintillometer methods, are commonly used as in-situ methods of measurement. They serve as standards to validate the observations made by UAVs or satellites. These techniques fall under the area of remote sensing and measure ETa in a contactless manner as well (Sathe & Mann, 2013). The eddy covariance (EC) method measures latent heat flux by analyzing the relationship between vertical wind velocity and moisture fluxes using high-fre□uency sensors. The ETa instruments include a 3-D wind sensor for measuring wind speed and sound, as well as an infrared gas analyzer to measure water vapor density and CO_2 concentration. The observation area typically spans hundreds to thousands of square meters. According to Vickers et al. (2010), the EC method's inaccuracy in predicting latent heat flux (LE) is approximately 5–20%. EC observation technology is complex due to intricate mathematical calculations and the lack of standardization across different observation sites. Several global flux observation networks, such as ChinaFLUX (China) and FLUXNET (Europe and North America), have been established, comprising numerous EC sites that have been crucial in collecting valuable data. Compared to EC, scintillometry is a relatively recent technology. Scintillometer instruments typically consist of a transmitter and a receiver. The receiver measures fluctuations in the transmitter's radiation caused by turbulent eddies in the atmosphere. These fluctuations are used to calculate sensible heat flux (H) and LE values on a landscape scale, typically spanning a few kilometers. Additionally, scintillometer instruments are highly effective on diverse surfaces and in challenging terrains (Veltink et al., 2005).

Scintillometers along with EC techniques have typically been employed as the benchmark to evaluate various techniques. Nevertheless, a recurring problem in these methods is the energy balance closure or lack thereof, where the combined sensible, ground, and latent heat fluxes do not match the net radiation. Several correction methods have been suggested to address the issue, and Wang et al. (2012) provided a review of three of these methods. According to one method, any errors in closure are believed to be caused by H, either underestimating or overestimating it, which leads to an imbalance issue. Meanwhile, LE is assumed to remain constant. In the other method, it is assumed that the closure errors can be ascribed to LE and the H value remains unchanged. In a different approach, researchers have utilized the Bowen ratio (BR) to separate the remaining energy balance into the following distinct components. Typically, it is necessary to correct the measured LE from ground instruments before applying the Bowen ratio approach. Currently, it is considered the most commonly utilized method for forcing closure. In addition, EC technique-based ET measurements often face significant gaps in observations. Numerous studies have put forth various methods to tackle this problem. For example, Falge et al. (2001) proposed the look-up tables; the marginal distribution sampling was suggested by Reichstein et al. (2005); machine learning-based methods were put forward by Bigaignon et al. (2020); and a multiple imputation method recommended by Hui et al. (2004). A recent study introduced a full-factorial scheme based on physics to fill the gaps in the EC-measured ET. Their method has been shown to outperform traditional gap-filling methods.

6.4. REMOTELY SENSED CROP WATER PRODUCTIVITY AT VARIOUS SCALES

Research on CWP has been summarized into three scenarios based on the spatial scale: (1) point scales; (2) field scales; and (3) regional scales. Table 6.2 showcases various findings from previous literature sources. This section will compare the accuracy performance of various remote sensing techniques applied in these three scenarios (Campos et al., 2018).

6.4.1. Ground Measurements-Based Point Scale

Eddy Covariance has the capability to measure both GPP values and the ETa values simultaneously. Nevertheless, the EC observation areas for EC can be influenced by various meteorological factors such as humidity and windspeed, as noted by Kljun et al. (2015). Measuring the cumulative GPP or ETa of the entire crop growth period over a specific area can be quite challenging. In a study conducted by Imukova et al. (2016), they compared the ETa measurements taken daily by the EC in a wheat farming land with those obtained through the method of soil-water balance. The findings revealed that the Eddy Covariance technique resulted in an overestimation of ETa by approximately 46% to 70%. In a study conducted by Ding et al. (2010), the accuracy of the Eddy Covariance method in estimating ETa was evaluated by assessing the results with a large-scale weighing lysimeter. The results showed that the Eddy Covariance method consistently underestimated ETa, with underestimations ranging from 4.8–21.8% during the daytime and 10.3–30.2% during the night-time.

It should be noted that the EC measurement can be influenced by factors such as meteorology and terrain, as highlighted by Cheng et al. (2021). However, it is important to note that the variations in both of these studies may be attributed to the use of different evaluation methods. When it comes to measuring GPP, there have been limited studies assessing the precision of EC methods, mainly because of the absence of effective measurement techniques. In a study conducted by Goulden et al. (1996), they examined the CO_2 flux using EC measurements and compared it with the chamber method in a northeastern US forest ecosystem. Their findings indicated a potential error range of – 0.3–0.8 $tC{\cdot}ha^{-1}{\cdot}year^{-1}$. In a study conducted by Baldocchi (2003), this method was also employed to assess the accuracy of measuring CO_2 flux on a global scale. The findings indicated that the margin of error was found to be approximately $\pm$ 0.5 $t{\cdot}ha^{-1}{\cdot}year^{-1}$.

By using Eddy Covariance, the GPP and ETa values can be measured simultaneously at a high frequency, allowing for the representation of CWP in the form of ecosystem water-use efficiency (WUEGPP, the ratio of GPP values to Eta values). As an illustration, Kuglitsch et al. (2008) assessed the daily and half-hour scale WUEGPP fluctuations in Europe measured by the EC. Similarly, Xie et al. (2016) investigated the response of Beijing, China's forest ecosystem to drought using EC-measured WUEGPP at various time scales. Ground measurements, such as EC methods, are being extensively utilized for estimating forest ecosystem CWP. In the agricultural sector, the high temporal resolution EC technique has the potential for application.

6.4.2. Field Scale-Dependent UAV Observations

In recent years, the use of UAV platforms for studying crop phenotyping has become increasingly prevalent (Jin et al., 2020). On the other hand, comparatively less research has been conducted on the CWP. UAVs equipped with thermal and multispectral sensors have become a popular tool for estimating ETa, as demonstrated by Wang et al. (2012). It is important to preprocess the observed data from UAVs (multispectral and thermal) before converting them into reflectance and temperature data. This involves tasks such as geometric and radiometric calibration; and image mosaic, which help to accurately interpret the images' digital number or DN values. Liu et al. (2021) provided a comprehensive analysis of these processes. In a study conducted by Hoffmann et al. (2016), two-surface energy balance-based models, namely the dual-temperature difference or the DTD; followed by the two-source energy balance or the TSEB model were employed to calculate a Barley crop's daily ETa. The findings revealed that the DTD model had an estimation accuracy of 26% (rRMSE) while the TSEB model had an estimation accuracy of 37% (rRMSE). In a study, the P-T model was utilized to estimate the LE values of a maize crop. The results were quite promising, with the R^2 value reaching 0.84 and the RMSE value being 41.19 $W \cdot m^{-2}$. In a recent study, the Mapping of Evapotranspiration at High Resolution with Internalized Calibration, or the METRIC model, specifically the SEBR variant, was employed to approximate the evapotranspiration or ETa of the almonds. The results showed a root mean square error (RMSE) of 1.23 $mm \cdot day^{-1}$. In a recent study, an unmanned aerial vehicle (UAV) was utilized to observe and estimate the crop coefficient (Kc) and subsequently determine the evapotranspiration (ETa) (Van et al., 2017). Machine learning algorithms were used to establish the relationship between VIs and Kc. The UAV observations provided an effective measurement of Kc, with an R^2 value of 0.65. There has been a growing interest in studying the estimation of field-scale crop evapotranspiration through UAV observation in recent years. Nevertheless, there is room for further improvement in the estimation accuracy (Mao et al., 2022).

When it comes to field-scale crop yields, SR is the main method used. In a recent study by Maimaitijiang et al. (2020), researchers utilized UAV observation data along with machine learning algorithms to calculate soybean yield. By analyzing structural, temperature, textural, and spectral information, they achieved impressive accuracy with an R^2 value of 0.72 and a rRMSE of 15.9%. In a study, it was found that there was a strong correlation between the multi-temporal phase Vegetation Indices obtained from UAV-based observations and the rice yield. The R^2 value was as high as 0.76, indicating a good agreement between the two variables (Meng & Cheng, 2020). Typically, indicators observed by UAVs that describe the growth status of crops have a strong correlation with CY. Thus, utilizing the SR approaches to investigate the correlation between yields is precise and logical. Nevertheless, it does not possess the necessary mechanism and portability, as pointed out in their recent study. In the assimilation of the CERES-wheat model, an RMSE value of 1008.08 $kg \cdot ha^{-1}$ and the rRMSE value reaching 9.36% were reported as the results. The hyperspectral indicators like leaf nitrogen content (LNC) and LAI observed by UAV were employed. The DA method has a more robust mechanism, although it does involve more intricate processes and parameters when compared to the Statistical Regression approaches. Irrespective of the approach, observations using UAVs possess the capability to accurately approximate crop yield on a field scale (Chu et al., 2017).

Directly estimating CWP is a challenging task for many studies. Although UAV observations are highly maneuverable, the weather makes it challenging to conduct continuous observations during the crop-growing phase. Just like studies that rely on ground measurements utilized UAV-based observations of the daily ETa and GPP to calculate WUEGPP. Their findings demonstrated a strong agreement with EC observations, with a strong R^2 value of 0.86 followed by an RMSE value of 0.35 $gC \cdot kg^{-1}$. Further research has examined the impact of, and drought monitoring on crop yield, but they did not provide a quantitative estimation of crop water productivity. Generally speaking, irrigation timing and field water management can be guided by UAV observations (Gonçalves et al., 2016).

6.4.3. Satellite Observations Based on Regional Scale

Satellite platforms can improve regional decision-making by providing wider coverage and continuous monitoring. In China, the performance of MOD16 and SEBAL products was evaluated by Cheng et al. (2021). The results indicated that the SEBAL-calculated ETa had an accuracy (RMSE) of 0.95 $mm \cdot day^{-1}$ in cropland, while MOD16 had an accuracy of 8.94 $mm \cdot 8\ days^{-1}$, based on validation using EC observations (Moulin et al., 1998). To estimate latent heat (LE) based on cropland EC using remote sensing (RS) indicators, six machine learning (ML) algorithms were assessed for their performance in China:

i. Long short-term memory (LSTM);
ii. Artificial neural network (ANN);
iii. Extreme gradient boosting (XGBoost);
iv. Support vector machine (SVM);
v. Random forest (RF);
vi. K-nearest neighbor (KNN).

The findings revealed that RMSE values ranged from 18.67 to 21.23 W·m^{-2}, while R^2 values ranged from 0.81 to 0.88. In another study, the performance of the MOD16 algorithm was analyzed in the US for describing cropland ETa. When Landsat data were used as input, the results showed an R^2 of 0.69 and an RMSE of 0.70 mm·day^{-1} when compared to EC-based ETa observations. Chen et al. (2021) evaluated the SEBS algorithm for describing global ETa using data from 238 sites, achieving an RMSE of 1.56 mm·day^{-1}.

Estimating crop yield at the regional scale can offer valuable insights for agricultural management as well as grain policy formulation. In a study, the authors utilized AquaCrop and incorporated RS data from the HJ-1A/B and the RADARSAT-2 to make estimations in Yangling, China on the winter wheat yield. The findings of the study demonstrated a favorable level of agreement with the actual measured yield, as indicated by an R^2 value of 0.42. Their study introduced a novel approach called spatial assimilation to approximate winter-wheat yields in the plains of Central North China. They combined the MCWLA-Wheat, a model that captures the relationship between weather and crops over a large area, with the measurement of LAI (Leaf Area Index). The results of their study showed promising performance, with a good R^2 value of 0.42 followed by an RMSE value of 562 kg·ha^{-1}. Generally, the methods of data assimilation are being used across various environments and crop circumstances due to their reliability and versatility.

Because of their effectiveness and simplicity, SR techniques have also been applied extensively. Noland et al. (2018), for instance, estimated alfalfa yield indirectly by utilizing several VIs computed from satellite-observed multispectral data. They discovered that these VIs could account for 81–90% of the yield fluctuation. In a recent study by Cao et al. (2021), they successfully utilized a combination of the enhanced vegetation index or the EVI and advanced deep-learning or DL algorithms to accurately approximate the yield of winter wheat in China. The results of their research were quite promising, with an R^2 value of 0.71. In the relatively latest study, the random forest algorithm was employed along with RS indicators like the meteorological data and the NDVI to accurately approximate the yields of three of the major crops: 1). Wheat; and 2). Rice; and finally, 3). Maize. The findings revealed an impressive R^2 value exceeding 0.56 and a remarkably low rRMSE of less than 18%. According to Blatchford et al. (2019), the DA technique is known for its robustness and universality, making it irreplaceable. Although the SR method may occasionally attain higher accuracy in calculating CY, it cannot match the strengths of the DA method (Cavanaugh et al., 2010).

At present, there is a limited availability of data products with high temporal and spatial resolution (yearly) from CY or CWP. In a recent study, high-resolution data from remote sensing and advanced modeling techniques were employed for the estimation of maize and wheat yields in China. The researchers utilized numerous remote sensing (RS) indicators along with random forest algorithms to achieve a spatial resolution of 1 km. An excellent result was achieved compared to the actual yield, with the value of rRMSE reaching 21.80% and 26.81% for yield and maize. In addition, it also integrated MOD16 data to create a Crop Water Productivity dataset for China, featuring a 1 km spatial resolution. However, regional CWP estimation is often overlooked in most studies. Currently, the main focus of research is on computing WUEGPP to understand the correlation between water and yield (Turner et al., 2018).

Table 6.2. Statistical Analysis of R^2 in Various Crop Biophysical Parameters

	Dry-B	Wet-B	LAI	Height	fAPAR	Yield	VWC
Mean	0.62	0.62	0.67	0.61	0.82	0.59	0.42
STD	0.23	0.22	0.20	0.27	0.18	0.24	0.18
Min	–0.07	0.01	0.00	0.01	0.15	0.00	0.04
1st quartile	0.46	0.49	0.57	0.40	0.77	0.45	0.31
Median	0.66	0.63	0.70	0.70	0.86	0.64	0.47
3rd Quartile	0.80	0.77	0.82	0.81	0.95	0.78	0.52
Max	0.993	0.990	0.998	0.990	0.980	0.990	0.837
Count	1240	288	1116	165	60	284	47

Source: Bahrami, H., McNairn, H., Mahdianpari, M., & Homayouni, S. (2022). A meta-analysis of remote sensing technologies and methodologies for crop characterization. Remote Sensing, 14(22), 5633. https://doi.org/10.3390/rs14225633.

6.5. *IN-SITU* MEASUREMENTS

Irrespective of the techniques employed for estimating CWP or conducting in-situ measurements, the determination of CWP is reliant on Crop Yields and Evapotranspiration of crops. Thus, the in-situ calculation of CWP is considered to be based on ETa and CY, respectively (Kennedy, 2014). There are two main ways to measure CY: through physical measurement or empirical estimations. Physical measurements are taken after harvest, including drying and weighing of both whole-plot and subplot harvests. According to Norman et al. (1995), whole-plot harvesting is the most precise technique and is typically used in the context of on-farm studies. Subplot harvesting is also somewhat accurate; however, it has the sensitivity to the size and geometry of the subplot chosen. Additional faults may be identified within one of the smaller subsections of the area. Therefore, there is a balance to be struck between the efficiency and accuracy of measurements.

Physical measurements are commonly used to validate the RS methods. Based on the knowledge of experts or farmers, empirical estimating is seen to be a useful technique for calculating yield. Research has indicated that farmers in certain countries, such as the Philippines and Nepal, may lack familiarity with their planting areas, leading to errors in crop yield estimation. In contrast, farmers in China and Indonesia, who are more ac□uainted with their fields, achieve higher accuracy in their estimations (Walker et al., 2004). It would be worth exploring whether the accuracy of empirical estimations can be improved by studying representative planting areas.

There are different approaches to measuring crop ETa, including plant physiology and hydrology methods. Water balance is used in hydrology methods to compute ETa, usually with the use of a lysimeter. These techniques are based on precise theories and have minimal margin for error. Studies conducted in the past have indicated that the calculation of ETa from a lysimeter can have an error range of 5 ~ 25%. This error is primarily attributed to inaccuracies in soil moisture measurements. In a study, the researchers examined six lysimeters using soils with very similar properties.

The findings revealed a small relative error, that is, 8%. Even this variation was primarily because of inconsistent practices in field management. In a study conducted by Evett et al. (2012), they examined the relationship between the lysimeter of an irrigated cotton field and the determined soil water balance. Their findings revealed a relative error ranging from 5% to 18%. Plant physiology methods involve the use of crop coefficient methods and sap-flow meters. Directly e□uipping a sap-flow meter onto a single plant allows for accurate measurement of plant transpiration.

It is important to consider that the calculations are done on just a single-plant scale, which means that errors are often due to the coverage of the canopy and not the actual instrument. In addition, Kc is commonly utilized when considering reference ETa. It is important to mention that the Kc is generally based on empirical data rather than being universally applicable. This is because there are still many factors that have not been fully taken into account, such as agronomic practice, soil type, variety, and climate. In a study, the conclusion they reached was that the estimated values of ETa based on Kc showed variations across different years. It is common practice to assess other approaches using empirical estimation and physical measurements. When it comes to UAV or satellite observations, the scintillometer and EC methods are widely recognized and commonly used as a benchmark by experts in the field (Sethuraman et al., 2010).

6.6. SUMMARY

The objective of this chapter was to provide a summary of the remote sensing methods commonly used for estimating crop water productivity (CWP). Additionally, it aimed to review the practical aspects of the estimation of CWP at different scales, including the point scale; field scale; and regional scales. The CWP is not estimated directly, rather it is calculated using the CY to ETa ratio. So, its estimation involves the concurrent estimations of ETa and CY. Remote sensing techniques such as ground measurements, SR, SEF, and SEBR can be used to assess ETa at various application scales. Remote sensing techniques, such as SR and DA methods, and ground measurements, can be used to gather information about CY at various application scales (Jayez & Masson, 2006).

When it comes to estimating ETa, it is important to consider the MOST theory. While it serves as the foundation for most energy balance models, its effectiveness using finer resolution models in complex terrain remains to be studied. To improve its mechanism and calibration, researchers suggest further advancements.

In addition, the calculation of sensible heat flux relies heavily on the aerodynamic roughness length. However, due to numerous uncertainties, it has become imperative to combine existing or develop a new approach to tackle this problem. One potential solution could be to combine it with an ML method, as suggested by Peng et al. (2022). Another area that requires further investigation is the up-scaling methods, which have not been thoroughly examined. There are limitations and uncertainties associated with various methods that rely on observing only once during a single day. In addition to enhancing the method, the utilization of satellites with high-frequency synchronization comes with the potential to enhance the accuracy of up-scaling and offer fresh perspectives for method refinement. Dealing with the issue of view angle effects in thermal-infrared observations is crucial for accurately estimating land surface temperature (LST), which is the key factor in calculating various energy fluxes. This is particularly important for ensuring the accuracy of evapotranspiration

(ET) estimations. In recent research, Jiang et al. (2022) have developed a comprehensive framework to address the angular effect of thermal-infrared remote sensing LST. Their approach has the potential to greatly reduce uncertainties in the estimations of ET when using single-source energy balance-based models. Additionally, Wang et al. (2023) have introduced an innovative angular normalization method to bring the temperature of the land surface to nadir from the global MODIS off-nadir. This groundbreaking work establishes a benchmark for LST products that are angle-consistent MODIS, which is crucial for accurate estimations of ET.

REFERENCES

1. Ahmed, B. M., Tanakamaru, H., & Tada, A. (2010). Application of remote sensing for estimating crop water requirements, yield, and water productivity of wheat in the Gezira Scheme. *International Journal of Remote Sensing, 31*(16), 4281–4294.
2. Arcucci, R., Zhu, J., Hu, S., & Guo, Y. K. (2021). Deep data assimilation: Integrating deep learning with data assimilation. *Applied Sciences, 11*(3), 1114.
3. Baldocchi, D. (2008). 'Breathing' of the terrestrial biosphere: Lessons learned from a global network of carbon dioxide flux measurement systems. *Australian Journal of Botany, 56*(1), 1–26.
4. Baldocchi, D. D. (2003). Assessing the eddy covariance technique for evaluating carbon dioxide exchange rates of ecosystems: Past, present and future. *Global Change Biology, 9*(4), 479–492.
5. Balzarolo, M., Anderson, K., Nichol, C., Rossini, M., Vescovo, L., Arriga, N., & Martín, M. P. (2011). Ground-based optical measurements at European flux sites: A review of methods, instruments, and current controversies. *Sensors, 11*(8), 7954–7981.
6. Bigaignon, L., Fieuzal, R., Delon, C., & Tallec, T. (2020). Combination of two methodologies, artificial neural network, and linear interpolation, to gap-fill daily nitrous oxide flux measurements. *Agricultural and Forest Meteorology, 291,* 108037.
7. Blatchford, M. L., Mannaerts, C. M., Zeng, Y., Nouri, H., & Karimi, P. (2019). Status of accuracy in remotely sensed and in-situ agricultural water productivity estimates: A review. *Remote Sensing of Environment, 234,* 111413.
8. Bouman, B. A. M. (1995). Crop modeling and remote sensing for yield prediction. *Netherlands Journal of Agricultural Science, 43*(2), 143–161.
9. Cabas, J., Weersink, A., & Olale, E. (2010). Crop yield response to economic, site, and climatic variables. *Climatic Change, 101*(3), 599–616.
10. Campos, I., Neale, C. M., Arkebauer, T. J., Suyker, A. E., & Gonçalves, I. Z. (2018). Water productivity and crop yield: A simplified remote sensing driven operational approach. *Agricultural and Forest Meteorology, 249,* 501–511.
11. Cao, C., Cai, Z., Xiao, X., Rao, J., Chen, J., Hu, N., & Xue, Y. (2021). The architecture of the SARS-CoV-2 RNA genome inside virion. *Nature Communications, 12*(1), 3917.

12. Carrassi, A., Bocquet, M., Hannart, A., & Ghil, M. (2017). Estimating model evidence using data assimilation. *Quarterly Journal of the Royal Meteorological Society, 143*(703), 866–880.

13. Castaldi, F.; Hueni, A.; Chabrillat, S.; Ward, K.; Buttafuoco, G.; Bomans, B.; Vreys, K.; Brell, M.; van Wesemael, B. Evaluating the capability of the Sentinel 2 data for soil organic carbon prediction in croplands. *ISPRS J. Photogramm. Remote Sens.* 2019, *147*, 267–282.

14. Cavanaugh, K. C., Siegel, D. A., Kinlan, B. P., & Reed, D. C. (2010). Scaling giant kelp field measurements to regional scales using satellite observations. *Marine Ecology Progress Series, 403,* 13–27.

15. Challinor, A. J., Watson, J., Lobell, D. B., Howden, S. M., Smith, D. R., & Chhetri, N. (2014). A meta-analysis of crop yield under climate change and adaptation. *Nature Climate Change, 4*(4), 287–291.

16. Chemura, A.; Mutanga, O.; Dube, T. Separability of coffee leaf rust infection levels with machine learning methods at Sentinel-2 MSI spectral resolutions. *Precis. Agric.* 2017, *18*, 859–881.

17. Cheng, B., Schwing, A., & Kirillov, A. (2021). Per-pixel classification is not all you need for semantic segmentation. *Advances in Neural Information Processing Systems, 34,* 17864–17875.

18. Cheng, C., Xie, P., Abdukerim, A., Chen, W., Chen, X., Chen, Y., & (PandaX-II Collaboration). (2021). Search for light dark matter–electron scattering in the PandaX-II Experiment. *Physical Review Letters, 126*(21), 211803.

19. Chiraz, M. C., Olfa, M. A. R. R. A. K. C. H. I., & Hamadi, H. A. B. A. I. E. B. (2022). Remote sensing and soil moisture data for water productivity determination. *Agricultural Water Management, 263,* 107482.

20. Chu, T., Starek, M. J., Brewer, M. J., Murray, S. C., & Pruter, L. S. (2017). Assessing lodging severity over an experimental maize (Zea mays L.) field using UAS images. *Remote Sensing, 9*(9), 923.

21. Clevers, J.G.P.W.; Kooistra, L.; van den Brande, M.M.M. Using Sentinel-2 data for retrieving LAI and leaf and canopy chlorophyll content of a potato crop. *Remote Sens.* 2017, *9*, 663.

22. Condon, A. G., Richards, R. A., Rebetzke, G. J., & Farquhar, G. (2002). Improving intrinsic water - use efficiency and crop yield. *Crop Science, 42*(1), 122–131.

23. Crago, R. D. (1996). Conservation and variability of the evaporative fraction during the daytime. *Journal of Hydrology, 180*(1-4), 173–194.

24. Davis, E.; Wang, C.; Dow, K. Comparing Sentinel-2 MSI and Landsat 8 OLI in soil salinity detection: A case study of agricultural lands in coastal North Carolina. *Int. J. Remote Sens.* 2019, *40*, 6134–6153.
25. Defourny, P.; Bontemps, S.; Bellemans, N.; Cara, C.; Dedieu, G.; Guzzonato, E.; Hagolle, O.; Inglada, J.; Nicola, L.; Rabaute, T.; et al. Near real-time agriculture monitoring at national scale at parcel resolution: Performance assessment of the Sen2-Agri automated system in various cropping systems around the world. Remote Sens. Environ. 2019, 221, 551–568.
26. Ding, S., Chi, M. M., Scull, B. P., Rigby, R., Schwerbrock, N. M., Magness, S., & Lund, P. K. (2010). High-fat diet: Bacteria interactions promote intestinal inflammation which precedes and correlates with obesity and insulin resistance in mouse. *PloS One, 5*(8), e12191.
27. Evett, S. R., Schwartz, R. C., Casanova, J. J., & Heng, L. K. (2012). Soil water sensing for water balance, ET, and WUE. *Agricultural Water Management, 104,* 1–9.
28. Falge, E., Baldocchi, D., Olson, R., Anthoni, P., Aubinet, M., Bernhofer, C., & Wofsy, S. (2001). Gap-filling strategies for defensible annual sums of net ecosystem exchange. *Agricultural and Forest Meteorology, 107*(1), 43–69.
29. Gómez, D.; Salvador, P.; Sanz, J.; Casanova, J.L. Potato Yield Prediction Using Machine Learning Techniques and Sentinel 2 Data. *Remote Sens.* 2019, *11*, 1745.
30. Gonçalves, J., Henri□ues, R., Alves, P., Sousa - Silva, R., Monteiro, A. T., Lomba, Â., & Honrado, J. (2016). Evaluating an unmanned aerial vehicle - based approach for assessing habitat extent and condition in fine - scale early successional mountain mosaics. *Applied Vegetation Science, 19*(1), 132–146.
31. Goulden, M. L., Munger, J. W., Fan, S. M., Daube, B. C., & Wofsy, S. C. (1996). Measurements of carbon sequestration by long - term eddy covariance: Methods and a critical evaluation of accuracy. *Global Change Biology, 2*(3), 169–182.
32. Hargreaves, G. H., & Samani, Z. A. (1982). Estimating potential evapotranspiration. *Journal of the Irrigation and Drainage Division, 108*(3), 225–230.
33. He, L.; Mostovoy, G. Cotton Yield Estimate Using Sentinel-2 Data and an Ecosystem Model over the Southern US. *Remote Sens.* 2019, *11*, 2000.
34. Hoffmann, H., Zhao, G., Asseng, S., Bindi, M., Biernath, C., Constantin, J., & Ewert, F. (2016). Impact of spatial soil and climate input data aggregation on regional yield simulations. *PloS One, 11*(4), e0151782.

35. Hosseini, M., Agereh, S. R., Khaledian, Y., Zoghalchali, H. J., Brevik, E. C., & Naeini, S. A. R. M. (2017). Comparison of multiple statistical techniques to predict soil phosphorus. *Applied Soil Ecology, 114,* 123–131.
36. Hui, C., Lee, C., & Rousseau, D. M. (2004). Psychological contract and organizational citizenship behavior in China: Investigating generalizability and instrumentality. *Journal of Applied Psychology, 89*(2), 311.
37. Imukova, K., Ingwersen, J., Hevart, M., & Streck, T. (2016). Energy balance closure on a winter wheat stand: Comparing the eddy covariance technique with the soil water balance method. *Biogeosciences, 13*(1), 63–75.
38. Jackson, J., & McKenzie, D. (1983). The geometrical evolution of normal fault systems. *Journal of Structural Geology, 5*(5), 471–482.
39. Jayez, J., & Masson, A. (2006). What room for viewpoints? In O. Bonami & Cabredo-Hofherr (Eds.), *Empirical issues in syntax and semantics* (Vol. 6, pp. 175–199).
40. Jin, S., Huang, M., Kwon, Y., Zhang, L., Li, B. W., Oh, S., & Ruoff, R. S. (2018). Colossal grain growth yields single-crystal metal foils by contact-free annealing. *Science, 362*(6418), 1021–1025.
41. Kang, Y., Khan, S., & Ma, X. (2009). Climate change impacts on crop yield, crop water productivity, and food security: A review. *Progress in Natural Science, 19*(12), 1665–1674.
42. Kawabata, M., Maeda, K., Yamanaka, M., Nakaoka, T., Kawabata, K. S., Adachi, R., & Sekiguchi, K. (2020). SN 2019ein: New insights into the similarities and diversity among high-velocity Type Ia supernovae. *The Astrophysical Journal, 893*(2), 143.
43. Kennedy, J. J. (2014). A review of uncertainty in in situ measurements and data sets of sea surface temperature. *Reviews of Geophysics, 52*(1), 1–32.
44. Kljun, N., Calanca, P., Rotach, M. W., & Schmid, H. P. (2015). A simple two-dimensional parameterization for flux footprint prediction (FFP). *Geoscientific Model Development, 8*(11), 3695–3713.
45. Kuglitsch, F. G., Reichstein, M., Beer, C., Carrara, A., Ceulemans, R., Granier, A., & Valentini, R. (2008). Characterization of ecosystem water-use efficiency of European forests from eddy covariance measurements. *Biogeosciences Discussions, 5*(6), 4481–4519.
46. Lambert, M.J.; Traoré, P.C.S.; Blaes, X.; Baret, P.; Defourny, P. Estimating smallholder crops production at village level from Sentinel-2 time series in Mali's cotton belt. Remote Sens. Environ. 2018, 216, 647–657.

47. Li, C., Liu, Z. Y., Fang, X. Y., & Guo, Y. B. (2018). Residual stress in metal additive manufacturing. *Procedia CIRP, 71,* 348–353.
48. Li, L., Tang, S., Zhang, Y., Deng, L., & Tian, Q. (2017). GLA: Global-local attention for image description. *IEEE Transactions on Multimedia, 20*(3), 726–737.
49. Liu, Y., Liu, J., Xia, H., Zhang, X., Fontes-Garfias, C. R., Swanson, K. A., & Shi, P. Y. (2021). Neutralizing activity of BNT162b2-elicited serum. *New England Journal of Medicine, 384*(15), 1466–1468.
50. Lobell, D. B., Cassman, K. G., & Field, C. B. (2009). Crop yield gaps: Their importance, magnitudes, and causes. *Annual Review of Environment and Resources, 34,* 179–204.
51. Maimaitijiang, M., Sagan, V., Sidike, P., Hartling, S., Esposito, F., & Fritschi, F. B. (2020). Soybean yield prediction from UAV using multimodal data fusion and deep learning. *Remote Sensing of Environment, 237,* 111599.
52. Mananze, S.; Pôças, I. Agricultural drought monitoring based on soil moisture derived from the optical trapezoid model in Mozambique. *J. Appl. Remote Sens.* 2019, *13*, 1–16.
53. Mao, P., Ding, J., Jiang, B., Qin, L., & Qiu, G. Y. (2022). How can UAV bridge the gap between ground and satellite observations for quantifying the biomass of desert shrub communities? *ISPRS Journal of Photogrammetry and Remote Sensing, 192,* 361–376.
54. McIntyre, G. A., Brooks, C., Compston, W., & Turek, A. (1966). The statistical assessment of Rb-Sr isochrons. *Journal of Geophysical Research, 71*(22), 5459–5468.
55. Meng, J., & Cheng, Z. (2020). Improving the estimation of soil-available nutrients at the sub-field scale using time-series UAV observations. *Remote Sensing Letters, 11*(8), 739–747.
56. Moulin, S., Bondeau, A., & Delecolle, R. (1998). Combining agricultural crop models and satellite observations: From field to regional scales. *International Journal of Remote Sensing, 19*(6), 1021–1036.
57. Noi, P.T.; Kappas, M. Comparison of random forest, k-nearest neighbor, and support vector machine classifiers for land cover classification using sentinel-2 imagery. *Sensors* 2018, *18*, 18.
58. Norman, J. M., Kustas, W. P., & Humes, K. S. (1995). Source approach for estimating soil and vegetation energy fluxes in observations of directional radiometric surface temperature. *Agricultural and Forest Meteorology, 77*(3-4), 263–293.

59. Peng, Y., Felce, S. L., Dong, D., Penkava, F., Mentzer, A. J., Yao, X., & Dong, T. (2022). An immunodominant NP105–113-B*07:02 cytotoxic T-cell response controls viral replication and is associated with less severe COVID-19 disease. *Nature Immunology, 23*(1), 50–61.
60. Pereira, L. S., Perrier, A., Allen, R. G., & Alves, I. (1999). Evapotranspiration: Concepts and future trends. *Journal of Irrigation and Drainage Engineering, 125*(2), 45–51.
61. Pla, M.; Bota, G.; Duane, A.; Balagué, J.; Curcó, A.; Gutiérrez, R.; Brotons, L. Calibrating Sentinel-2 Imagery with Multispectral UAV Derived Information to Quantify Damages in Mediterranean Rice Crops Caused by Western Swamphen (*Porphyrio porphyrio*). *Drones* 2019, *3*, 45.
62. Prime, M. B., Sebring, R. J., Edwards, J. M., Hughes, D. J., & Webster, P. J. (2004). Laser surface-contouring and spline data-smoothing for residual stress measurement. *Experimental Mechanics, 44,* 176–184.
63. Reichstein, M., Falge, E., Baldocchi, D., Papale, D., Aubinet, M., Berbigier, P., & Valentini, R. (2005). On the separation of net ecosystem exchange into assimilation and ecosystem respiration: Review and improved algorithm. *Global Change Biology, 11*(9), 1424–1439.
64. Rozenstein, O.; Haymann, N.; Kaplan, G.; Tanny, J. Validation of the cotton crop coefficient estimation model based on Sentinel-2 imagery and eddy covariance measurements. *Agric. Water Manag.* 2019, *223*, 105715.
65. Sathe, A., & Mann, J. (2013). A review of turbulence measurements using ground-based wind lidars. *Atmospheric Measurement Techniques, 6*(11), 3147–3167.
66. Sethuraman, V. A., Chon, M. J., Shimshak, M., Srinivasan, V., & Guduru, P. R. (2010). In situ measurements of stress evolution in silicon thin films during electrochemical lithiation and delithiation. *Journal of Power Sources, 195*(15), 5062–5066.
67. Shuttleworth, W. J., Gurney, R. J., Hsu, A. Y., & Ormsby, J. P. (1989). FIFE: The variation in energy partition at surface flux sites. *IAHS Publ, 186*(6), 523–534.
68. Sibley, M. H., Kuriyan, A. B., Evans, S. W., Waxmonsky, J. G., & Smith, B. H. (2014). Pharmacological and psychosocial treatments for adolescents with ADHD: An updated systematic review of the literature. *Clinical Psychology Review, 34*(3), 218–232.
69. Söderström, M.; Piikki, K.; Stenberg, M.; Stadig, H.; Martinsson, J. Producing nitrogen (N) uptake maps in winter wheat by combining proximal crop measurements with Sentinel-2 and DMC satellite images

in a decision support system for farmers. *Acta Agric. Scand. Sect. B Soil Plant Sci.* 2017, *67*, 637–650.

70. Sun, N.; Wang, P.; Huang, F.; Li, B. Developing an integrated index based on phenological metrics for evaluating cadmium stress in rice using Sentinel-2 data. *J. Appl. Remote Sens.* 2018, *12*, 1.
71. Turner, A. J., Jacob, D. J., Benmergui, J., Brandman, J., White, L., & Randles, C. A. (2018). Assessing the capability of different satellite observing configurations to resolve the distribution of methane emissions at kilometer scales. *Atmospheric Chemistry and Physics, 18*(11), 8265–8278.
72. van der Meij, B., Kooistra, L., Suomalainen, J., Barel, J. M., & De Deyn, G. B. (2017). Remote sensing of plant trait responses to field-based plant-soil feedback using UAV-based optical sensors. *Biogeosciences, 14*(3), 733–749.
73. Vanino, S.; Nino, P.; De Michele, C.; Falanga Bolognesi, S.; D'Urso, G.; Di Bene, C.; Pennelli, B.; Vuolo, F.; Farina, R.; Pulighe, G.; et al. Capability of Sentinel-2 data for estimating maximum evapotranspiration and irrigation requirements for tomato crop in Central Italy. *Remote Sens. Environ.* 2018, *215*, 452–470.
74. Veltink, P. H., Liedtke, C., Droog, E., & van der Kooij, H. (2005). Ambulatory measurement of ground reaction forces. *IEEE Transactions on Neural Systems and Rehabilitation Engineering, 13*(3), 423–427.
75. Vickers, D., Göckede, M., & Law, B. (2010). Uncertainty estimates for 1-h averaged turbulence fluxes of carbon dioxide, latent heat, and sensible heat. *Tellus B: Chemical and Physical Meteorology, 62*(2), 87–99.
76. Vuolo, F.; Neuwirth, M.; Immitzer, M.; Atzberger, C.; Ng, W.T. How much does multi-temporal Sentinel-2 data improve crop type classification? *Int. J. Appl. Earth Obs. Geoinf.* 2018, *72*, 122–130.
77. Vyas, S. S., Nigam, R., Bhattacharya, B. K., & Kumar, P. (2016). Development of real-time reference evapotranspiration at the regional scale using satellite-based observations. *International Journal of Remote Sensing, 37*(24), 6108–6126.
78. Walker, J. P., Willgoose, G. R., & Kalma, J. D. (2004). In situ measurement of soil moisture: A comparison of techniques. *Journal of Hydrology, 293*(1-4), 85–99.
79. Walter, N., Cody, M. J., & Ball-Rokeach, S. J. (2018). The ebb and flow of communication research: Seven decades of publication trends and research priorities. *Journal of Communication, 68*(2), 424–440.

80. Wang, K. C., Dickinson, R. E., Su, L., & Trenberth, K. E. (2012). Contrasting trends of mass and optical properties of aerosols over the Northern Hemisphere from 1992 to 2011. *Atmospheric Chemistry and Physics, 12*(19), 9387–9398.
81. Wei, J., Alfajaro, M. M., DeWeirdt, P. C., Hanna, R. E., Lu-Culligan, W. J., Cai, W. L., & Wilen, C. B. (2021). Genome-wide CRISPR screens reveal host factors critical for SARS-CoV-2 infection. *Cell, 184*(1), 76–91.
82. Weiss, M., Jacob, F., & Duveiller, G. (2020). Remote sensing for agricultural applications: A meta-review. *Remote Sensing of Environment, 236,* 111402.
83. West, H.; Quinn, N.; Horswell, M.; White, P. Assessing vegetation response to soil moisture fluctuation under extreme drought using sentinel-2. *Water* 2018, *10*, 838.
84. Wikle, C. K., & Berliner, L. M. (2007). A Bayesian tutorial for data assimilation. *Physica D: Nonlinear Phenomena, 230*(1-2), 1–16.
85. Xiang, Y. T., Yang, Y., Li, W., Zhang, L., Zhang, Q., Cheung, T., & Ng, C. H. (2020). Timely mental health care for the 2019 novel coronavirus outbreak is urgently needed. *The Lancet Psychiatry, 7*(3), 228–229.
86. Yuan, Y., Liu, L., Chen, H., Wang, Y., Xu, Y., Mao, H., & Liang, H. (2016). Comprehensive characterization of molecular differences in cancer between male and female patients. *Cancer Cell, 29*(5), 711–722.
87. Zeman, S. (2019). The emergence of viewpoints in multiple perspective constructions. *Pragmatics, 29*(2), 226–249.
88. Zhang, Y., Li, X., Liu, X., & Qiao, J. (2011). The CA model is based on data assimilation. *Journal of Remote Sensing, 15*(3), 475–491.
89. Zhang, Z.; Liu, M.; Liu, X.; Zhou, G. A new vegetation index based on multitemporal sentinel-2 images for discriminating heavy metal stress levels in rice. *Sensors* 2018, *18*, 2172.
90. Žížala, D.; Juřicová, A.; Zádorová, T.; Zelenková, K.; Minařík, R. Mapping soil degradation using remote sensing data and ancillary data: South-East Moravia, Czech Republic. *Eur. J. Remote Sens.* 2019, *52*, 108–122.

Chapter 7

REMOTE SENSING OF RICE CROP AREAS

Contents

7.1. INTRODUCTION

Remote sensing is a key method and technology used to observe, map, and monitor the ecosystems that grow rice over large areas and over long periods of time to understand them from a number of different angles. More than half of the population of the world depends on rice as a critical staple diet, with most of them in developing nations in Latin America, Africa, and Asia (Fairhurst & Dobermann, 2002). As a result, rice is crucial to food security. Global production of paddy rice in 2011 was 720 million tons, with Asia accounting for almost 90% of that total. It was expected to slightly grow in 2012 (Figure 7.1).

Food security is questionable, nevertheless, as global production of rice has been slowing down recently (FAO, 2004c), the yearly consumption of rice has surpassed its production. The food demand of the world is predicted to rise in tandem with the expanding population, with projections of almost 9 billion people by the year 2050 (Shen et al., 2009).

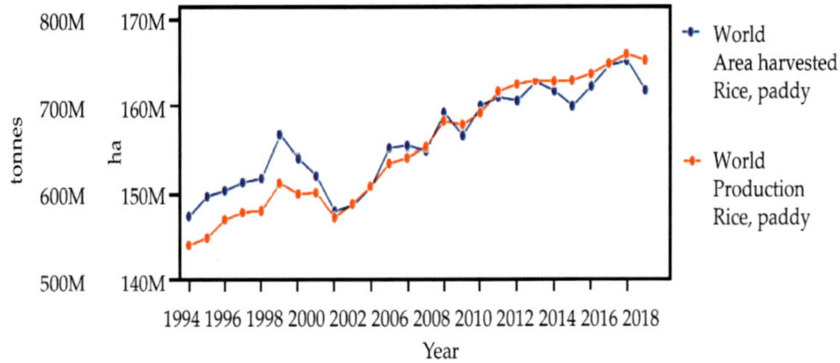

Figure 7.1. *Global production of Pady rice (million tons).*

Source: Mohidem, N. A., Hashim, N., Shamsudin, R., & Che Man, H. (2022). Rice for food security: Revisiting its production, diversity, rice milling process, and nutrient content. Agriculture, 12(6), 741. https://doi.org/10.3390/agriculture12060741.

To assure enhanced and sustainable agricultural output with improved yield, novel planting techniques and an expansion of the rice paddies are required (Sakamoto et al., 2009). However, the goal of increased food security based on rice agriculture may be hampered by the declining amount of water for irrigation and the arable land, along with the global climate change effects.

Effective methods of monitoring the cultivation of rice and the associated social, economic, and global environmental aspects are desperately needed to evaluate the rice production and demand equilibrium and to offer and enhance information on the growth conditions and ecosystems of rice production. Remote sensing provides fast, affordable, and precise quantitative information about the current state of the crops as well as a range of characteristics of crop biochemistry and physiology. Information produced from remote sensing is a valuable resource for decision-making processes related to multi-annual agricultural developments (like technical or cropping system advancements), as well as seasonal responses that are strictly linked to the rice growth status. Standardized approaches provide a viable alternative to time-consuming and expensive conventional field research for evaluating the geographical and temporal environmental changes that can impact the health and growth of crops (Faisal et al., 2012).

i. Remote sensing can give accurate and up-to-date information for many applications connected to rice farming systems, such as.
ii. Monitoring and mapping an eye on the size of rice-producing ecosystems.
iii. Assessing and monitoring the health and growth status of rice.
iv. Evaluating the effectiveness of cropping systems and patterns.
v. Calculating the metrics associated with crop growth, such as biomass, evapotranspiration, fraction of absorbed photosynthetically active radiation or fAPAR, and LAI.
vi. Precision agriculture, which includes prediction of yields, applying pesticides and fertilizers for vi. Sustainable agriculture and differentiating crops from weeds.
vii. Input, enhancement, and expansion of models for agricultural yield and growth.
viii. Assessment of the agricultural ecosystems alongside the evaluation of the adverse environmental effects, like soil salinity and land degradation.
ix. A deeper knowledge of how the paddy crops interact with their surroundings.
x. Hydrologic modeling and assessment (soil moisture, field productivity, water consumption monitoring, and the connection between hydrological processes and crop yield).
xi. Disease incidence measurements (pests and drought).
xii. Estimating methane fluxes relevant to climate change emissions (Usali & Ismail, 2010).

Through various activities and approaches, along with the use of GIS and other relevant data, a solid foundation has been established for making informed decisions in areas of cropping optimization, farm management, farming system intensification, sustainable irrigation, and food provision management policy (Hedley et al., 2016).

7.2. THE SPATIAL DISTRIBUTION OF RICE-GROWING AREAS

Rice is grown throughout a broad range of landscape types within the huge spatial span between 80° longitude and 60° latitude (Xiao et al., 2006). It is present in various regions across the globe, including sub-tropical and tropical areas in Asia; parts of Africa (like the Valley and Nile Delta in Egypt, Sudan, and Madagascar); Central Asia; the Middle East (including Iran and Iraq); temperate European regions (such as Spain and Italy), certain states in the USA (such as Louisiana, Missouri, and Mississippi) as shown in Figure 7.2.

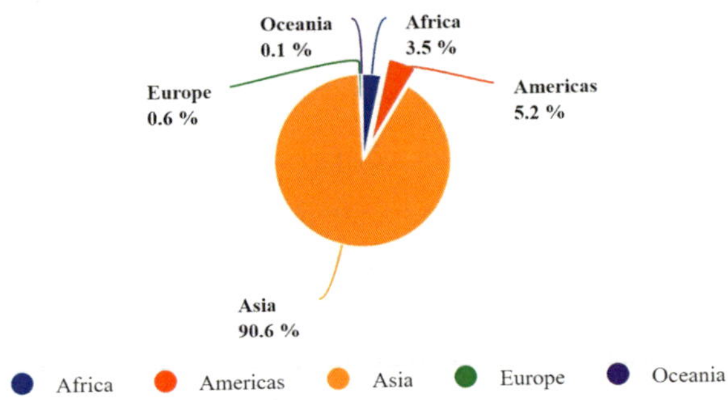

Figure 7.2. *Representation of global rice-growing regions.*

Source: Mohidem, N. A., Hashim, N., Shamsudin, R., & Che Man, H. (2022). Rice for food security: Revisiting its production, diversity, rice milling process, and nutrient content. Agriculture, 12(6), 741. https://doi.org/10.3390/agriculture12060741.

Despite the wide distribution of growing areas, the cultivation of rice re□uires specific ecological conditions. The yield and distribution of rice are influenced by various factors, as determined by Shao et al. (2001). Temperature plays the most crucial role in the growth of rice. To produce just one rice, crop each year in a region, the accumulated annual temperature, measured in degree days needs to be between 2000 and 4500°C for ≥10°C.

To cultivate two rice crops in a year, the temperature should be between 4500 and 7000°C for ≥10°C. For the maximum production of three rice crops a year, the temperature should be 7000°C for ≥10°C. Understanding the distribution and yield of specific rice varieties relies on various factors such as wind speeds in the growing stages; relative humidity; and lengths of days. According to research conducted by Shao et al. (2001), rice plants thrive in soils that have certain characteristics.

These include clayey or sandy soils with near-neutral pH values; good permeability; great fertility and high water-retention capacity. Factors such as soil permeability; evaporation; irrigation; and rainfall play a crucial role in determining water availability and supply. The rice-cultivating areas typically consist of wetland ecosystems, which re□uire significant irrigation or precipitation.

Crucial water supplies provided through the seasonal precipitation in the sub-tropics and the tropics in Southeast Asia and the Indian subcontinent are influenced by the dominant monsoon climate system. Understanding the timing of cropping seasons in irrigated areas is influenced by the alternating dry and wet seasons of the monsoon system, as well as the amount of water available. This information was highlighted by Jiang et al. (2011).

7.3. RICE PLANT CHARACTERISTICS, THE GROWTH CYCLE, AND THE AGRICULTURAL SYSTEM

Understanding the growth stages and physiology of rice plants is important for uses and the planning phases of projects involving RS. Therefore, the following text will provide a concise overview of rice plant characteristics, their growth cycles, and potential farming systems.

Rice is classified under the Oryza genus, Oryzeae tribe, and Gramineae family (De Datta 1981). Two of such significant species cultivated are 1). *Oryza glaberrima*, the common species that originated in West Africa; and 2). *Oryza sativa* L., which has roots in the Asian continent. The basic structure includes culms, roots, and leaves that form tillers. As the rice seeds approach maturity, they give rise to an inflorescence in the shape of a panicle. The three growth phases of rice are classified by the International Rice Research Institute.

- The vegetative stage is from germination until the start of panicles.
- The reproductive phase, spans from the initiation of the panicle to flowering.
- The ripening stage occurs between flowering and grain maturity.

The growing cycle consists of three growth phases, each with 10 growth stages. These stages are: (i) germination stage; (ii) seedling stage; (iii) tillering stage; (iv) elongation of stem stage; (v) panicle booting/initiation stage; (vi) heading stage; (vii) flowering stage; (viii) milk stage; (ix) dough stage; and finally (x) mature grain; the different stages of the crop growth cycle are categorized into three distinct phases: vegetative (0–3); reproductive (4–6); and ripening (7–90) (IRRI, 2009a; Figure 7.3).

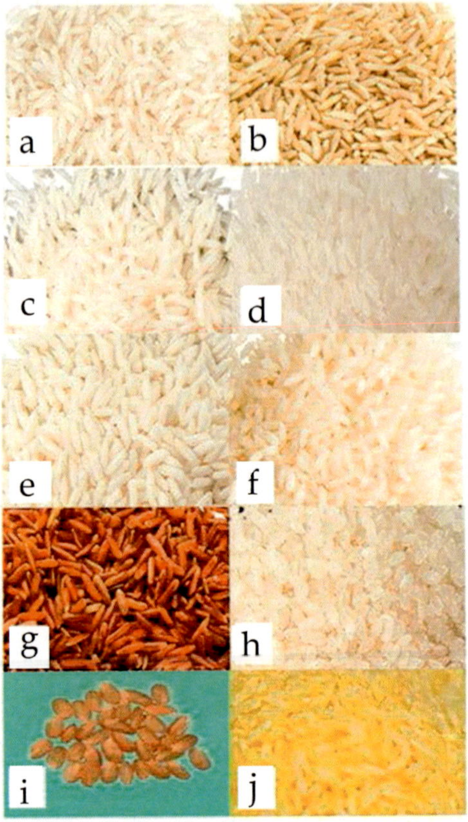

Figure 7.3. *The various varieties of rice: (a) white rice; (b) brown rice; (c) basmati; (d) fragrant rice; (e) glutinous rice; (f) Ponni rice; (g) red rice; (h) japonica rice; (i) weedy or red rice; and (j) golden rice.*

Source: Mohidem, N. A., Hashim, N., Shamsudin, R., & Che Man, H. (2022). Rice for food security: Revisiting its production, diversity, rice milling process, and nutrient content. Agriculture, 12(6), 741. https://doi.org/10.3390/agriculture12060741.

The length of the growth cycle is determined by the specific species of rice alongside the prevailing climatic circumstances, and it is primarily influenced by the duration of the vegetative phase. In tropical regions, the reproductive phase lasts approximately 35 days, followed by a ripening phase, lasting around 30 days (IRRI, 2009b). In temperate areas, the rice varieties typically have a life cycle of 140–150 days, while in tropical regions, this duration drops to about 110–120 days. Ultimately, the life cycle of the plant spans 3–6 months, varying based on the specific cultivar and the surrounding environment. In colder and temperate tropical climates, a single crop is cultivated annually, with a longer cropping duration. Nevertheless, it has been observed that yields tend to be greater. Rice harvesting can occur 2–3 times a year in tropical regions with ideal irrigation. Crop calendars and growing seasons can vary depending on the variety of crops. The crop cycle initiation is determined through the water distribution scheme, as discussed by Le Toan et al. (1997). Based on the situation, the crop calendar may vary among adjacent groups. For instance, Le Toan et al. (1997) discovered a significant variation of about 6 to 8 weeks in the timing of fields in one specific region in Indonesia.

According to Lampayan (2015), rice is a unique crop that thrives in a wetland ecosystem. Thus, the ecosystems that support rice can be categorized into the four primary farming systems, taking into account factors such as topography; type of soil; temperature; drainage; and water regime (Figure 7.4).

To begin with, dryland or upland rice is cultivated in plateaus or rainfed highlands. According to research conducted by Bambaradeniya & Amarasinghe (2003), the rice is dry seeded. This is primarily because of the low humidity, inadequate soil quality, and the absence of water accumulation on the surface or in the rhizosphere. The yields are typically quite low, averaging close to 1 t ha^{-1}. Upland rice farming is prevalent in the regions of Southeast Asia; India; Madagascar; and Brazil. Upland rice production accounts for roughly 13% of the total rice-planted area worldwide and contributes about 4% to global rice production.

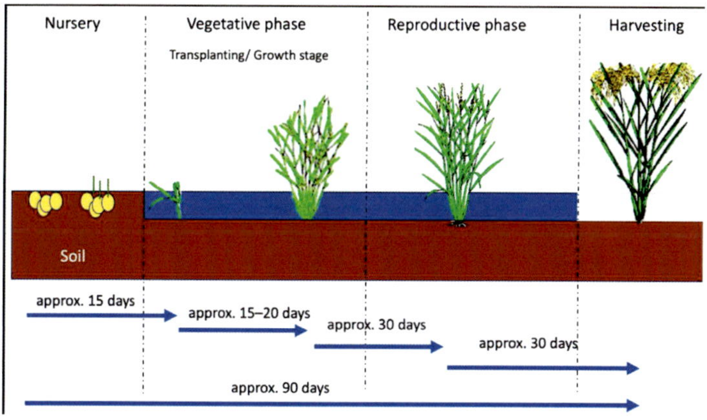

Figure 7.4. *Representation of rice land ecosystems.*

Source: Fernández-Urrutia, M., Arbelo, M., & Gil, A. (2023). Identification of paddy croplands and its stages using remote sensors: A systematic review. Sensors, 23(15), 6932. https://doi.org/10.3390/s23156932.

Additionally, lowland rainfed rice ecosystems experience intermittent flooding with varying depths (1–50 cm) as well as durations. The presence of water in the bounded fields is influenced by the intensity of rainfall, as noted by Bambaradeniya & Amarasinghe (2003). Lowland rice fields typically show anaerobic soil conditions and have stagnant water for over one-fifth of the crop period. Just 25% of the world's rice acreage and about 17% of its production are grown in such systems, and it is primarily located in Madagascar and Africa. The average per-hectare yield is approximately 2.5 tons.

Irrigated lowland rice is the third type of farming system. It is primarily farmed in Asia in confined flooded fields, with pre-germinated seeds planted in moist seedbeds. Ensuring that pond irrigation is consistently available for at least two-thirds of the duration of the crop is crucial for such rice crops. The water presence throughout the growing season contributes to the high productivity of rice. The water depth is maintained within a range of 5–10 cm. During the rainy season, the yield remains around 5 t ha^{-1}. When it comes to irrigated rice, it is possible to have 1–3 harvests per year. Thus, this farming system holds immense global significance. This irrigated rice makes up over 50% of global rice cultivation, with two-thirds of the global production of rice coming from these irrigated areas.

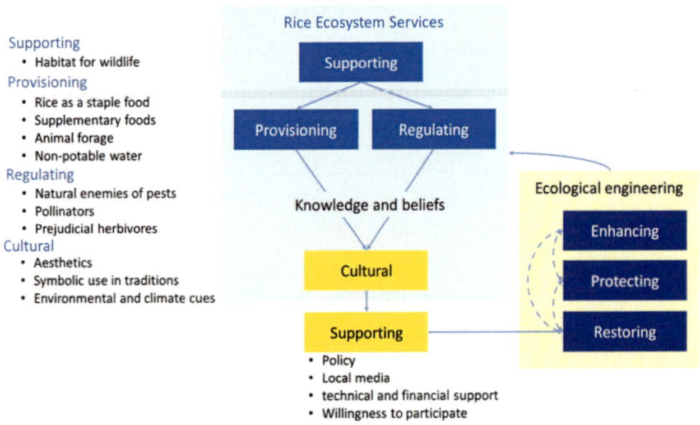

Figure 7.5. *The linkage between the ecosystem services provided by tropical rice fields (indicated in the □lue frame) and 'human services to the rice ecosystem' (indicated in the yellow frame) as mediated through policy, media, and culture (supporting services to ecosystems (indicated by an orange rectangle)). The schematic presents the linkages as part of a positive feedback system largely driven by farmers' appreciation of ecosystem services (knowledge and □eliefs). Some of the supporting, provisioning, regulating, and cultural services mentioned by farmers in this study are also listed.*

Source: Horgan, F. G., Vu, Q., Mundaca, E. A., & Crisol-Martínez, E. (2022). Restoration of rice ecosystem services: 'Ecological engineering for pest management' incentives and practices in the Mekong Delta region of Vietnam. Agronomy, 12(5), 1042. https://doi.org/10.3390/agronomy12051042.

The final farming systems are generally located in flood-prone wetland farming areas with water depths ranging from 0.5–3 m; they receive water from lakes, rivers, or tides in case of river mouth deltas. Such systems are present in South America, West Africa, and Southeast Asia. Productivity is often low due to the climate-related hazards of floods and droughts. This system plays a crucial role for small farms located in flood-prone areas, where the yield is limited to 1 t ha^{-1}. Irrigated rice has a yearly productivity that is 5 times higher compared to rainfed rice, 12 times higher than deep-water rice, and up to 100 times higher than upland rice (Fairhurst & Dobermann, 2002). Wetland rice is considered a significantly crucial farming system, serving as the foundation for numerous research works and studies. It is common to refer to rice fields as paddies or paddy fields in these studies. The name's original root is padi, a Malayan word meaning rice. In Southeast Asia, Figure 7.5 provides an insightful view of the geographical distribution of rice-cropping ecosystems.

7.4. EFFECTS OF RICE-GROWING ECOSYSTEMS ON THE ENVIRONMENT

Rice-growing regions are widely recognized as some of the environmentally sound; most sustainable; and efficient agroecosystems in the world. The transient aquatic agroecosystem seen in paddy rice fields is comprised of both biotic and non-biotic elements and is defined by a cohesive habitat. Biodiversity is a crucial factor in the success of an agroecosystem, as it has an essential role in the different services provided by the ecosystem. Loss or changes in species composition or richness led to a loss or reduction of other functions or species that are closely linked. This can upset or even wreck the agro-ecosystem balance (Heong, 2008).

A tropical rice-growing wetland ecosystem provides valuable services that have a direct impact on people (IRRI, 2006). Nevertheless, ecosystems based on rice are susceptible to artificial impacts. Excessive use of fertilizers and pesticides can diminish the capacity to provide services and goods. For instance, in tropical rice-growing environments, one of the most crucial functions is the suppression of pests (like planthoppers). Using insecticides in tropical rice is not necessary and can be counterproductive as they interfere with the natural/ingrained pest suppression abilities of the plant. Unfortunately, in recent times, the widespread use of insecticides with broad coverage and lasting residual effects has negatively impacted the natural pest control systems in rice cultivation. Rice insecticides have been identified as highly toxic agrochemicals and had a significant share of the crop insecticide global markets in 1988 (Gumma et al., 2011).

Pesticides and fertilizers have frequently been used in a prophylactical manner, which increases the number of pollutants in surface and ground waterways. When fertilizers and pesticides are improperly applied or are over-applied, they cause long-lasting and severe damage to ecosystems that grow rice. Through the food chain and water use, nitrogen contamination; soil deterioration; eutrophication; and groundwater pollution have a direct effect on human health. In addition, the persistent presence of water can lead to the spread of diseases. This creates breeding grounds for mosquitoes, which are the main vectors for diseases such as schistosomiasis, dengue, and malaria. These diseases can have an impact on rice farmers (Tan et al., 2019).

Fertilizer overuse is related to the fact that nitrogen has an impact on crop productivity and is crucial for stabilizing and enhancing yield production and crop growth. The N status before the heading stage has a direct correlation with rice yield. For optimal rice yield and quality, n-topdressing must occur during the panicle start stage. Thus, evaluating the canopy nitrogen using RS methods during these stages of growth can offer valuable insights for the application of nitrogen fertilization. Rice yield and growth can be significantly affected by climate changes and impacts. These include global warming, which can lead to shifts in temperature, including night-time temperatures rise. Intense weather events like typhoons, cyclones, and storms; changes in rainfall patterns; unusually longer growing seasons; droughts (particularly in rainfed systems); and flooding, also pose risks to rice crops (Che Salmah et al., 2017).

7.5. WATER FUNCTIONS IN RICE-GROWING ECOSYSTEMS

Wetland rice is a widely practiced farming method for the global production of rice, resulting in significant water consumption. According to Bouman (2009), approximately 2500 liters of water is required through irrigation and/or rainfall to yield/cultivate just 1 kg of rough rice in the rice field. According to Bouman (2009), 859 cubic kilometers of water are used annually by evapotranspiration on all rice fields globally. Rice irrigation uses between 25% and 33% of the freshwater resources globally. With the rise in population; growing industrial and urban needs; the depletion of resources; and pollution, fresh water is becoming an increasingly valuable and scarce natural resource. Therefore, there is a strong demand for implementing water-saving technologies and enhancing water resource management. This includes the adoption of aerobic varieties of rice and the implementation of novel irrigation techniques and regimes. Specifically, the practice of multi-cropping paddy rice farming hinders the ecosystem's ability to dry adequately. This is caused by the continuous and rapid flooding of the rice fields (Krupa et al., 2011). Soil undergoes various negative effects on its chemical and biological processes, such as waterlogging, which results in a slowed rate of humus decomposition, reduced ammonia volatilization, reduced soil nitrogen mineralization, salinity build-up, and the accumulation of soluble iron and manganese. Anaerobic wetland soils also release methane into the atmosphere, contributing to increased greenhouse gas emissions. Methane (CH_4) accounts for 5–10% of global greenhouse gas emissions, making it the second most significant greenhouse gas after CO_2. It is produced as a result of anaerobic fermentation of soil organic matter. Given methane's significant global warming potential—28 times that of CO_2—it is essential to account for methane emissions from flooded rice fields when calculating the global carbon balance (Tuong & Bouman, 2003).

7.6. RICE REFLECTANCE FEATURES IN REMOTE SENSING DATA

Efficient and timely monitoring and mapping of paddy rice are crucial for assessing environmental and agricultural productivity, analyzing water and food security, and studying greenhouse gas (GHG) emissions, among other important factors. Agronomic and ecological factors can support this effort by relating to the backscatter or spectral properties of Earth observation data, which may provide indirect information on plant development and growth. Understanding the phenological development of crops is essential for effective crop monitoring, as it offers valuable insights into the current condition of cultivated crop species and their interactions with environmental factors (Kuenzer & Knauer, 2013).

A rice crop canopy's reflectance spectrum is the result of complex interactions between its biochemical and biophysical characteristics. Changes in growth conditions, such as variations in climate or the supply of nutrients or water, can significantly impact the reflection patterns and characteristics. However, when these conditions are similar, different rice types tend to exhibit a comparable temporal spectral response when considering optical data (Darvishsefat et al., 2011).

7.6.1. Rice Reflectance Features in Optical Remote Sensing

During the initial few weeks following transplantation, the ground cover percentage is less than 15% while the rice seedlings heal from the trauma of the transplantation. The reflectance characteristics are primarily influenced by the underlying soil and water surface. During the active tillering (6th) stage, which occurs 9 weeks after transplanting, there is a rapid increase in the number of tillers and leaf area. The reflectance in the visible section reaches its lowest value after 12 weeks of transplanting, while the NIR region achieves its maximum value of the season. The red edge inflection point shows a strong correlation with the leaf chlorophyll content (Van Niel & McVicar, 2004).

Leaf pigments, including carotenoids and chlorophyll a and b, have a significant impact on the absorption of incoming visible radiation (Figure 7.6) The physiological state of the plant, which ultimately influences a plant's productivity, is closely correlated with the levels of these substances. Understanding photosynthetic functioning is crucial for predicting and diagnosing plant growth and estimating the exchange of carbon GHG involving the atmosphere and the ecosystems. There are clear variations in the foliar chemical content throughout the senescence and ripening stages, which can indicate physiological stress. The red reflectance of a rice crop can be seen as an inversely proportional indicator of its greenness (Duan et al., 2019). The red reflectance decreases from the initial 10% during the emergence phase to just 2% during the flowering stage, followed by rising again to 16%–18% at maturity.

This change is caused by the loss of green brightness in the stems and leaves, as well as the yellowing of the grains of rice. Over time, the NIR reflectance shows variations that correspond to changes in biomass. It starts at a minimum value of 15% during the early tillering phase and reaches a peak value of 50% at the heading stage. Afterward, it gradually decreases as the biomass decreases from the loss and decay of leaves, reaching its last value of around 33%. Figure 7.7 illustrates this progression in multispectral data.

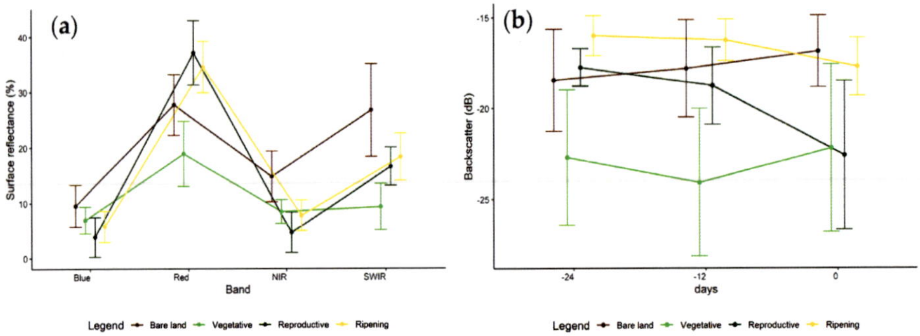

Figure 7.6. *(a) Spectral □and reflectance of PROBA-V related to rice growth stages; (b) backscattering of VH on three consecutive dates.*

Source: Ramadhani, F., Pullanagari, R., Kereszturi, G., & Procter, J. (2021). Mapping cloud-free rice growth stages using the integration of PROBA-V and Sentinel-1 and its temporal correlation with sub-district statistics. Remote Sensing, 13(8), 1498. https://doi.org/10.3390/rs13081498.

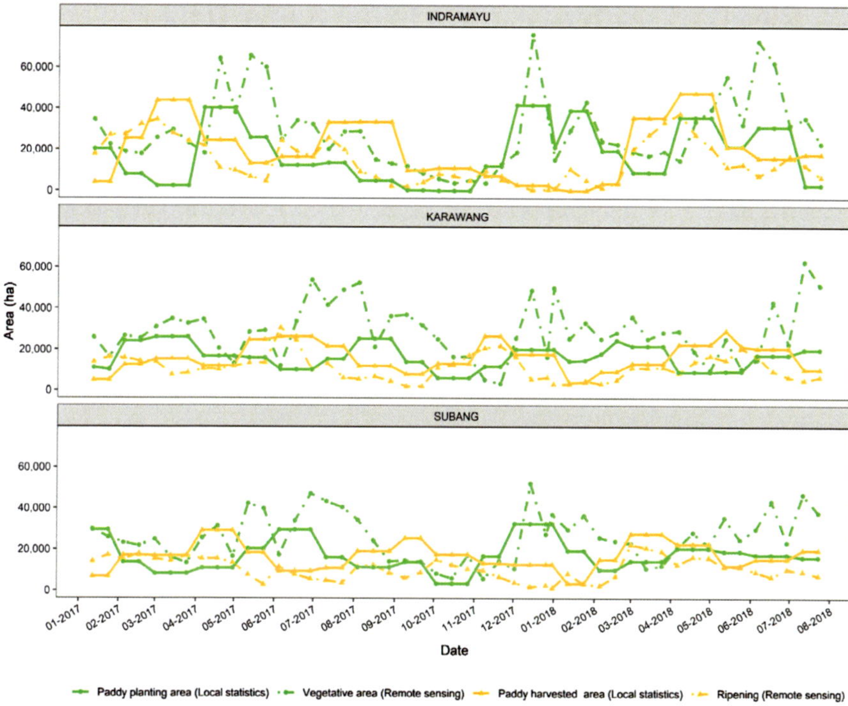

Figure 7.7. *The comparison of temporal rice growth stages (vegetative and ripening) with local statistics (planting and harvested area) for Indramayu, Karawang, and Subang Regency from January 1, 2017 to July 23, 2018.*

Source: Ramadhani, F., Pullanagari, R., Kereszturi, G., & Procter, J. (2021). Mapping cloud-free rice growth stages using the integration of PROBA-V and Sentinel-1 and its temporal correlation with sub-district statistics. Remote Sensing, 13(8), 1498. https://doi.org/10.3390/rs13081498 (Creative Commons License).

In most situations, the water encompassing the rice plant has an impact on its spectrum in the actual world. The water regime system significantly influences the canopy reflectance spectra, as demonstrated by Xue et al. (2005). The visible range reflectance, that is, 460–710 nm is higher in intermittent irrigation systems compared to shallow water systems in the mid-infrared range, that is, 1480–1650 nm. However, it is lower in the NIR region, that is, 760–1220 nm) for the shallow water system. An intermittent irrigation system has lower biomass; shoot numbers; and LAI than a shallow water system. This leads to increased reflection across the layers of leaves and, consequently, a lower NIR spectrum reflectance. In the intermittent irrigation system, the leaf structure tends to be compact with a lower leaf chlorophyll content. Thus, in the visible bands, there is somewhat higher reflectance and lower absorption (Mosleh et al., 2015).

7.6.2. Rice Backscatter Features in the Radar Data

Having a deep understanding of the synthetic aperture radar or SAR backscatter of rice fields in relation to crop growth is crucial to developing robust and reliable methods for extracting crop growth parameters and crop phenology information. In the section that follows, the overall mechanism of rice backscatter as well as the temporal dynamics of backscatter for various incident angles, polarizations, and frequencies, during the growing season are discussed. Additionally, a relationship between the growth parameters, polarization, and frequency is depicted (Ribbes, 1999).

Radar backscatter from crops is very receptive to the canopy structure alongside the underlying soil surface. There is a significant sensitivity when it comes to crops that have elongated, vertical and unique canopy elements, like the ones found in cereals such as ears, leaves, and stems. The presence of water and a fluctuating amount of vegetation-covered land define the underlying surface.

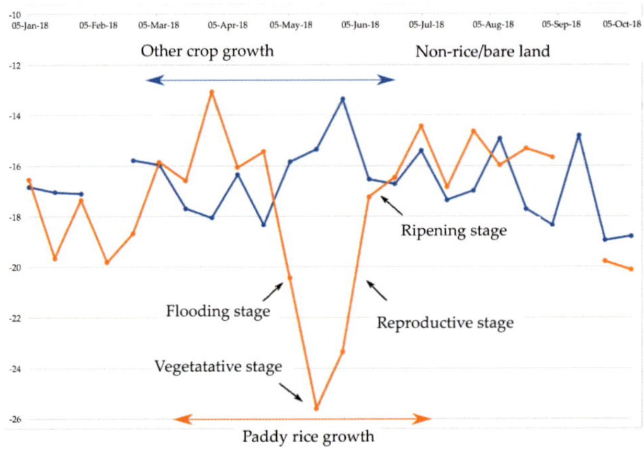

Figure 7.8. *The example of rice detection workflow for each pixel with vertical-horizontal backscattering (VH) values. The blue line is the example of a non-rice cultivation class (–7° 10' 34.9", 111° 52' 55.42"), and the orange line is the example of a rice cultivation class (–6° 27' 57.27", 108° 8' 56.64").*

Source: Ramadhani, F., Pullanagari, R., Kereszturi, G., & Procter, J. (2020). Automatic mapping of rice growth stages using the integration of Sentinel-2, MOD13Q1, and Sentinel-1. Remote Sensing, 12(21), 3613. https://doi.org/10.3390/rs12213613.

As a result, the primary scattering mechanism occurs due to the interaction between radar waves, vegetation, soil, and water. When it comes to backscatter, the most important factors are usually the structure of the plant; the gravimetric water content; plant biomass per unit volume; and plant height. Figure 7.8

illustrates the primary scattering mechanisms involving direct scattering (Ze & Shihua, 2023). Each frequency band typically exhibits distinct patterns in their backscatter signatures, which are influenced by the rice canopy growth. Typically, there is a significant variation in backscatter levels throughout the different stages of rice crop growth, from pre-planting to pre-harvesting. Before transplanting, the radar signal recorded is significantly low because of the specular reflection caused by the flooded fields.

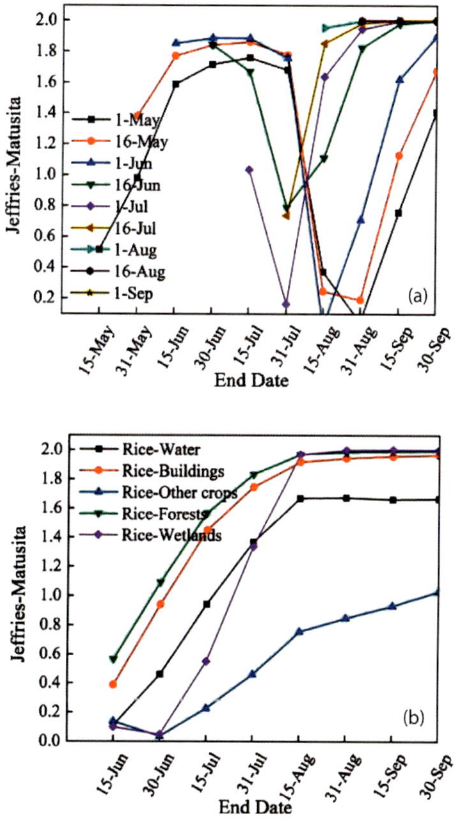

Figure 7.9. *JM distance illustrates different characteristics of various land cover types. (a) Shows the JM of the VH-sum feature across different times for various dates. (b) Represents the JM distance across rice water, rice buildings, rice-other crops, rice forests, and rice wetlands for different dates. Image (b) demonstrates the JM distance of the VH-slope feature in SART2.*

Source: Wang, G., Meng, D., Chen, R., Yang, G., Wang, L., Jin, H., Ge, X., & Feng, H. (2024). Automatic rice early-season mapping based on simple non-iterative clustering and multi-source remote sensing images. Remote Sensing, 16(2), 277. https://doi.org/10.3390/rs16020277.

During this stage, it is possible to distinguish rice fields from various non-rice areas in radar images due to the noticeable difference in backscatter between non-flooded and flooded regions. Having access to flooded field data is crucial to accurately identify rice fields using multi-temporal radar data. During the vegetation phase, there is an increase in backscatter due to volume scattering within the rice canopy and multiple reflections across the water and plant and water surfaces. This increase reaches its maximum at the heading stage, as observed by Chen & McNairn (2006). The leaf and stem densities along with plant water content both decline with crop ripening. Throughout the ripening stage till harvest, there is a decrease in backscatter coefficient, the stems continue to be the predominant source of total backscattering coefficient across all polarizations, even if the grain densities are increasing at this point in the rice's growth. Rice crops are the only ones with this temporal backscatter pattern. Backscatter variations throughout the rice growing season are significantly greater compared to other crops in agriculture. Figure 7.9 provides a visual representation of this (Dong et al., 2006). The selected wavelength, polarization, and incidence angles all affect the backscattered signal, regardless of the rice crop's stage of growth. When it comes to polarization, it is worth noting that certain widely applicable interrelations can be considered. In the early stages of rice growth, HH-polarized backscattering coefficients tend to be lower than VV-polarized backscattering coefficients. This is due to the physical structure of the rice plant, which is characterized by short vertical leaves and stems during the early stages of growth. However, throughout most of the rice-growing season, HH-polarized backscattering coefficients are consistently higher than VV-polarized ones because the vertical structure of the rice plants causes attenuation of the waves. Therefore, utilizing multi-angular and multi-polarization data provides valuable insights for monitoring rice growth throughout the season. In addition to polarization, the choice of incidence angles and radar fre□uency also has a significant impact.

The backscattering factors change more quickly at smaller incident angles for the Ku and Ka bands, while they change more rapidly at larger incident angles for the L and C bands (Inoue et al., 2002). HH-polarized signals approach saturation at the heading point in the C-band, where the double-bounce vegetation–water mechanism is the predominant scattering mechanism for both HH and VV polarization. This could be due to a reduction in water and changes in the physical size of the rice during the harvest season. According to HH-polarized data in the L-band domain, there is relatively little backscatter during the initial flooding period, followed by rising backscatter throughout the vegetative development period, and a leveling off during the ripening/heading stage. The increase in backscatter during the growth period aligns with the increase in biomass, attributed to the dominant vegetation-water interaction scattering mechanism (Le Toan et al., 1997). In another study by Le Toan et al. (1997), the backscatter coefficient ($\sigma°$) was examined in relation to rice parameters using VV-polarized data from the ERS-1 SAR C-band. The study found that during the productive phase, $\sigma°$ values ranged from –8 to –6 dB and remained consistent until the end of the growth cycle. They highlighted that the significant temporal variation in radar backscatter is a notable characteristic of rice fields, distinguishing them from other types of land cover (Table 7.1).

Table 7.1. *Challenges to the Sustainability of the Rice Production System and Suggested Proposed Solutions*

Challenges to the Sustainability of the Rice Production System
• Increasing food prices;
• Depletion of the soil's nutrient pool;
• Degradation of soil health;
• Decline in soil structure;
• Depletion of groundwater supplies;
• Rising production costs;
• Labor deficit;
• Burning of crop waste;
• Greenhouse gas emissions;
• Climate change vulnerability;
• Resistance to herbicides in weeds;
• Crop reaction to applied stressors has decreased;
• A diverse weed flora;
• Increasing labor cost;
• Issues with residue management;
• Abiotic stresses such as drought, salinity, flood, storm, and very low and high temperatures;
• Rice grain quality.
How Can We Overcome the Challenges?
• Agricultural diversification;
• Conservation agriculture;
• Direct seeded rice;
• Dry-wet method of irrigation in rice;
• Furrow irrigated raised bed (FIRB) system of planting;
• Automated irrigation systems;
• Early sowing of zero-till wheat;
• Sustainable residue/waste management practices;
• Precision nutrient management;
• Short or medium season rice varieties;
• Success in capacity building of the stakeholders;
• Integrated weed management with herbicide rotation.

Source: Mohidem, N. A., Hashim, N., Shamsudin, R., & Che Man, H. (2022). Rice for food security: Revisiting its production, diversity, rice milling process, and nutrient content. Agriculture, 12(6), 741. https://doi.org/10.3390/agriculture12060741.

Kim et al. (2008) observations of the temporal fluctuations in backscattering coefficients at various polarizations and incidence angles across different bands are documented. Inoue et al. (2002) discovered that the VV polarization in the L-band is not very sensitive at more steep incident angles such as 25° or 35°, but exhibits more noticeable seasonal changes at larger angles like 45° and 55°. During the ripening period, a reversal in the backscattering coefficient in the X-band was observed with the incident angle. It decreased at 25° and increased at 55°.

However, no comparable phenomenon occurred in the Ku and Ka bands. Furthermore, the X-band response can be explained by factors such as canopy structure and penetration depth. Their investigations demonstrated that microwave backscatter is extremely sensitive to the size of dominant elements, as evidenced by the seasonal trends observed at the higher frequencies, that is, the X, Ku, and Ka bands. Canopy scattering has a strong influence on higher frequencies, making them particularly responsive to early growth changes, especially during the month after transplanting when LAI is less than 1.5. On the other hand, lower frequencies such as the L and P bands are affected by both canopy and soil backscatter, contributing significantly to the overall backscatter. Using complex and distinctive signatures, SAR backscatter may be linked to the biomass of rice crops. This information can then be utilized to map important parameters for the production of rice and ultimately estimate crop yield (Oyoshi et al., 2016).

In addition to the radar radiation characteristics, the rice field structure, such as planting density, also has a significant impact. For instance, due to varying plant spacing, which causes both destructive and constructive interference between rice plants, there can be a significant variance in the scattered return signals in the L-band from rice fields with diverse architectures. Understanding the plant spacing is crucial for utilizing the L-band scattering results to accurately determine the rice biomass. It was also discovered that the C-band is less affected by the structure of rice fields compared to the L-band, with a difference of about 6 cm versus 24 cm. In contrast found that row spacing significantly impacted the backscatter of radar. In general, there is inconsistency among the specific parameters determined by different studies, as well as the strength of correlations and the degrees of significance. The backscatter signals from vegetated surfaces are influenced by numerous factors, such as roughness; soil moisture; stand structure; and plant biomass. These factors interact with sensor configurations, including incident angles; polarizations; and frequency. Due to the complexity of these interactions, it is not always possible to derive statements that are universally valid in all cases.

7.7. REMOTE SENSING OF RICE-GROWING AREAS

Extensive research has focused on evaluating ecophysiological and crop yield factors using remote sensing techniques across various wavelengths, including microwave, thermal, and optical. These methods are often combined with modeling approaches to enhance accuracy and understanding. Assessing the biophysical variables and typical characteristics of crops in rice-growing areas involves deriving data from remote sensing, allowing for a comprehensive evaluation of the following factors:

- Wet biomass (WBM).
- Above-ground dry biomass (DM).
- Leaf area index (LAI), an essential factor in determining forecasting yield, crop growth dynamics, and estimating foliage cover.
- Plant height.
- Green leaf chlorophyll density (GLCD), crucial in precision agriculture. It indicates the photosynthesis activity and nutritional nitrogen levels in the green vegetation and helps measure crop responses to nitrogen application.
- Crop yield, a crucial factor for rural development and an indicator of national food security. Several studies (Patel et al., 1985) have highlighted its significance. Bastiaansen and Ali (2003) also evaluated the importance of initiating food trade promptly to meet national demand. The first step toward achieving agricultural production forecasting is to forecast the crop area on an annual basis (Fang et al., 1998).
- Photosynthetically active radiation (PAR) that is absorbed (APAR) and fractioned (fPAR). According to Bastiaansen and Ali (2003), the PAR value represents the total radiation that can be used for photosynthesis if leaves capture all of it. It is important to note that solar radiation is transmitted and reflected by leaves. As a result, only a portion of photosynthetically active radiation (PAR) is absorbed by the canopy, known as absorbed PAR (APAR), which is crucial for the assimilation of carbon dioxide. Research has demonstrated a correlation between the fraction of PAR absorbed by a plant canopy (fPAR) and net primary productivity (NPP). This relationship is determined by an efficiency coefficient that quantifies the amount of fixed carbon per unit of intercepted radiation (Gandharum et al., 2021).
- Water content of plants.

- Straw and final grain.
- Treatment with nitrogen.
- Rice cultivar; plant chlorophyll content; water background color, clarity, and depth.

One of the more sophisticated uses of remote sensing data involves the modeling of crop growth. Such models replicate the biophysical processes throughout the whole crop cycle, taking into account as many soil, climate, and other factors as possible to give a complete picture of how plants develop and grow.

The accuracy of the model heavily relies on the chosen set of variables, as it is developed based on a functional relationship for specific variables in a specific environment. Therefore, it is important to consider multiple biophysical processes when developing models to accurately predict crop yield. By utilizing input parameters obtained from remote sensing, the model parameters can be permanently calibrated, ensuring spatial integrity. The following is the function of crop models in conjunction with remote sensing (Islam, 2021).

1. They enable a non-destructive; fast; alongside cost-effective assessment of the status of crops.
2. They allow for the utilization of model outputs, such as LAI and ground cover, in the models of growth simulation.
3. They enable spatial extrapolation with satellite imagery to the regional level.
4. They ensure that model results are repeatable and objective (Mansaray et al., 2017).

7.7.1. Rice Field Remote Sensing Using Optical Data with High to Medium Resolution

Medium-resolution satellite sensors are crucial for distinguishing paddy fields. Acquiring data is quite affordable and the rice field's resolution is adequate for distinguishing between them. As a result, it is feasible to capture the entire cultivation area using just a few scenes. This section covers around 30 publications with subject areas spanning 13 countries (Frolking et al., 2002).

The data used in this study were obtained from various satellite sources, including Landsat, SPOT, and IRS. The spatial resolution of the data ranged from 20 meters from the SPOT sensors to 188 meters from the IRS Wide-Field Scanner (WiFS). Utilizing data from MOS-1, a Japanese satellite, that has a spatial resolution of 50 m, is an exceptional application (Ribbes, 1999).

7.7.1.1. Application Fields

One of the primary uses of medium-resolution optical data is to identify and map paddy fields. McCloy et al. (1987) carried out one of the earliest investigations on this subject. In Australia's New South Wales, they monitored rice-growing regions for two years using data from Landsat MSS. In a study conducted by Turner & Congalton (1998), rice fields located in the Inner Niger Delta, Mali were successfully mapped using three SPOT-XS images from a single year. They encountered challenges due to the low vegetation cover outside and within the cropped region; resulting in region heterogeneity; and the small sizes of the cropped fields. Ishiguro et al. (1993) utilized MOS-1 and Landsat Thematic Mapper data to estimate the rice area in a specific region of Japan. There was some confusion due to the shadowy tree canopies and small sizes of fields, which led to some overlap with the forested and city areas.

Panigrahy, Ray, and Panigrahy carried out a comprehensive land-cover classification with an emphasis on rice (2009). They successfully implemented a crop classification for the state of Orissa (Cuttack district), India, using data from the advanced wide-field sensor (AWiFS) on the IRS-P6 satellite. The classification included vegetables, ground nuts, and rice. Change detection is another typical use for optical data with a medium resolution. This tool can be incredibly valuable for measuring and monitoring trends in paddy fields, including the trends in composition and extent; paddy field state resulting from crop rotation; damage from floods or storms; or natural transformation of vegetation. These can be identified by analyzing multiple images of a single area taken at different seasons or years and comparing or subtracting the images using mathematical techniques. In two separate studies, conducted by Bailey et al. (2001), researchers analyzed land-cover change using ETM+ Landsat data and compared it to digitized maps. Bailey et al. specifically examined the size of digitized fields of wild rice in Minnesota, USA, in 1994 using aerial photographs. They then overlaid this information with a 1999 Landsat ETM+ classification and their findings revealed a significant 65% crop loss, which was attributed to subse□uent flooding and increased precipitation of the potential fields. In a study conducted by Son and Tu (2008), they analyzed the conversion of forestland in the lower Mekong Delta, Vietnam. They compared the data from 2001 ETM+ with a land-use map from 2005. They also found a significant decrease in forestland, with a large portion being converted into irrigated rice fields.

Additional research was conducted to evaluate the impact of floods and other calamities on the paddy fields. For instance, a study conducted by Okamoto, Yamakawa, and Kawashima in 1998 examined the extent of flood damage to rice production in North Korea during 1995. Through the use of change detection among multiple Landsat TM imagery, it was determined that a significant portion of the rice paddies in North Korea were impacted by the floods, affecting approximately 42% of the total area. The flood irrigation used in rice fields throughout the growing seasons creates optimal breeding grounds for mosquito vectors like malaria. In a study conducted by Diuk-Wasser et al. (2006), Landsat ETM+ data was utilized to detect and track breeding grounds of malaria-carrying mosquitoes in a rice-growing region in Mali.

Based on IRS-P6 AWiFS data, Badarinath et al. (2001) examined the burning of crop residue and the ensuing greenhouse gas emissions in the paddy fields on the Indo-Gangetic Plains, India. Based on the analysis of burnt area classifications and subse□uent estimation of emissions, it was found that GHG emission levels from rice residue had been somewhat higher compared to wheat residues. Therefore, it is recommended to incorporate the residues into the soil for better management.

Estimating rice yields is another way to utilize optical satellite data with medium resolution. In a study conducted by Patel et al. (1991), they utilized IRS 1-A LISS-I data along with the transformed Landsat MSS data to examine the correlation between the yield data and the spectral reflectance in two districts of the state of Orissa, India. In a study conducted in Thailand, Tennakoon et al. (1992) used Landsat TM images to identify paddy fields. They then correlated the reflectance values from the images with yield information gathered through interviews with farmers. Strong correlations were found and a model for estimating crop yield was developed.

7.7.1.2. Different Methods

Given the range of data and study areas, there is a wide variety of processing methods that are employed. Various approaches are used for digitization and classification, depending on the satellite data selected as the foundation. These methods can involve manual techniques or utilize VIs, combinations or bands, or a single band. However, no decision-tree classification; object-based methods, or other complex methods based on medium-resolution satellite data were available.

One widely used method on a regional level involves visually interpreting and digitizing rice fields on a computer screen. This can take the form of visually selecting training regions for later supervised classification, digitizing field borders fully by hand, or visually analyzing spectral data to determine the threshold.

In addition, successful applications of both unsupervised and supervised classifications have been made for mapping paddy fields. For example,

Fang (2001) applied unsupervised classification to Landsat-5 TM data from China's Hubei Province. Initially, the unsupervised classification was performed, followed by the derivation of 10 major land cover categories through comparison with various thematic maps. This method demonstrated satisfactory accuracies, achieving an impressive 90% for early rice and 81% for semi-late rice. Supervised classifications based on ground truth training data are more commonly used, typically resulting in higher accuracies. In a rice-growing region of Mali, Diuk-Wasser et al. (2006) used a supervised maximum likelihood classification (MLC) on Landsat ETM+ scenes. By utilizing a dataset with 12 bands from two scenes captured during the growth and harvest periods, the accuracy of rice delineation reached an impressive 98%. The significance of bands 3 and 4 (NDVI) and bands 5–7 (middle infrared) was emphasized. MLC is widely used for identifying paddy fields using medium-resolution remote sensing data.

Apart from utilizing multi-date data, prior image transformation is fre□uently done for the input data to enhance classification outcomes. To optimize input data sets, Panigrahy & Sharma (1997) modified LISS-I data using principal component transformation. Panigrahy et al. (2009) studied this further using AWiFS data. Bailey et al. (2001) used a different transformation called the minimum noise fraction transformation. This data was obtained via a Landsat ETM+ multi-date data set to enhance the identification process for wet vegetation (Cao et al., 1996).

One area of research that holds great significance is the utilization of various vegetation indices, including MSI (moisture stress index); LAI; EVI; and NDVI. In a study conducted by Oguro et al. (2001), the researchers explored the application of EVI and NDVI data from various sensors in combination with SAR data to effectively monitor rice fields. Based on their analysis, it was determined that utilizing a combination of EVI and NDVI is the most effective approach. This is due to the ability of these two metrics to differentiate between forests and rice fields, resulting in an impressively low error rate of less than 2%.

In addition, Panigrahy et al. (2009) conducted tests on various indicators including MSI, NDVI, and a self-developed three-band index (TBI) composed of SWIR, NIR, and red band to differentiate and classify crops. Overall, it was found that including the SWIR band, which indicates water stress, in a vegetation index for classifications improved the overall accuracy. Using the depth of ETM+ band 5 (SWIR); MSI; and the normalized difference infrared index (NDII) as a reference, Van Niel et al. (2003) examined the rice classification results with NDVI and a standard MLC of all six reflective ETM+ bands, with an emphasis on the utility of moisture-based indices. This study highlights the significance of considering the environmental moisture in the rice fields' classification process. Additionally, it emphasizes the importance of selecting the appropriate index based on the specific stage of the growing cycle (Bresler & Dagan, 1983).

7.7.2. Rice Field Remote Sensing Using Optical Data with Low-Resolution

Various available sensors like SPOT VEGETATION; MODIS; and AVHRR have lower spatial resolutions but offer higher temporal resolution and wider swath width. This allows for cost-effective coverage of large areas compared to sensors with higher spatial resolution. Having a high temporal resolution is crucial for analyzing agricultural systems as it makes it possible to capture the dynamic growth of crops throughout the year. Another reason to consider using these low spatial resolution data is the large scale of the research area, which can span continents or even the entire globe. This is made possible by the wide swath widths of roughly 2200, 2330, and 2500 km SPOT VEGETATION, MODIS, and AVHRR, respectively. These large-scale studies often center on the shifting patterns of land cover worldwide and the potential connections between the climate; the biogeochemical cycles; human activities; and finally, these patterns (Motohka et al., 2009).

7.7.2.1. Fields of Application

Every 1–2 days, the MODIS sensor on board the Terra and Aqua satellites observes the entire Earth's surface using 36 spectral bands at spatial resolutions of 250, 500, and 1,000 meters. The available data includes a range of scientific standard products, such as vegetation indices (VIs). Therefore, the sensor is a valuable tool for obtaining comprehensive measurements of land surface conditions over extended periods of time.

As experts in crop management and remote sensing, Chen, Son, and Chang (2012) conducted an analysis in the Mekong Delta, specifically examining shifts in rice-cropping systems. They utilized a MODIS NDVI time series from 2001 to 2007, with a 250 m spatial resolution. Their rice monitoring procedure involved three steps: first, they used empirical mode decomposition (EMD) to filter noise; next, they performed endmember extraction, which was used to train a linear mixture model for classification. Finally, the results were validated by comparing them with statistical data and ground truth. Their classification achieved an overall accuracy of 90.1%.

In a study conducted by Gumma et al. (2011), an 8-day composite MODIS NDVI time series, spanning a period of 10 years (2000–2009) with a 500 m spatial resolution, was used to monitor changes in a rice-growing region in Nepal. The study began by using Shuttle Radar Topography Mission (SRTM) data to divide the study area based on the steepness of the gradients. Next, the application of an isodata classification algorithm yielded a range of classes, with the number varying based on the segment. Additional data, including high-resolution Google Earth images and ground truth data, were used to identify the classes. Ultimately, the unsupervised classes were combined into a total of 11 classes. The resulting maps were evaluated by comparing them to field plot data and Nepal's district-level rice area statistics, with accuracy varying from 67% to 91% for different rice classes. Based on these maps, the rice cultivated area experienced a 13% decline in 2006 due to a severe drought.

The EVI and NDVI, two MODIS products are generated globally in the aforementioned geographic resolutions at intervals of 16 days. NDVI is well-known for its sensitivity to chlorophyll and its ability to detect shifts in the visible portion of the spectrum. On the other hand, EVI is more sensitive to near-infrared and is particularly effective in areas with high biomass. As such, there is a strong complementarity between the two indices. As previously mentioned, VIs are crucial for monitoring the growth of rice. MODIS VI products are highly sought after in vegetation research, especially for monitoring rice phenology over a long period of time (Verbano & Venturini, 2011).

This also holds for the sensors SPOT VEGETATION and AVHRR. In a study conducted by Singh et al. (2006), the researchers utilized NDVI data from AVHRR to examine changes in the timing of rice growth, specifically in terms of peak vegetation and emergence. During the period from 1981–2000, there was a noticeable shift in the rice-growing pattern in Punjab, India, with the planting season starting 3–4 weeks earlier. The Special Sensor Microwave Imager or SSM/I, a passive microwave radiometer, was also utilized because the AVHRR NDVI data were not sensitive enough at the time of crop emergence. Using a time series of VEGETATION NDVI data along with rainfall data, were able to differentiate between rainfed and irrigated paddy rice fields in the Suphanburi province of Thailand, specifically concerning SPOT VEGETATION. Accurate mapping of the irrigated areas plays a crucial role in enhancing water management, particularly in quantifying water usage. They analyzed the peak NDVI in relation to the mean rainfall data. The results revealed strong relationships between a single peak of NDVI and peak rainfall for rainfed rice and multiple peaks of NDVI for irrigated rice. Consequently, there were weaker correlations with rainfall (Smith & Oatley, 1955). In addition to utilizing VIs, the utilization of low spatial resolution sensors can be valuable in mapping significant land-cover classes. For instance, Panigrahy et al. (2005) utilized an unsupervised classification method to analyze agricultural areas using SPOT

VEGETATION data. Their objective was to identify cropping patterns and crop rotation practices. In the Indian state of Orissa, cropping patterns were mapped over three distinct seasons, from which the rotation of crops was developed. Therefore, it is worth noting that a comprehensive quality assessment has not yet been conducted, but seven different crop rotation methods and four distinct crop classes have been identified (Berth & Jeschke, 1989).

7.7.2.2. Methods

The wavelet-based filter for determining crop phenology, or WFCP, is a novel technique developed to detect phenological phases of rice crops from MODIS data. The WFCP comprises three procedures. The first involves preliminary processing of satellite data, opting for the Enhanced Vegetation Index (EVI) due to its improved atmospheric correction, as highlighted by Huete et al. (2002). This choice is particularly crucial as the rice-growing season overlaps with the rainy season, characterized by elevated humidity levels. The index was created using the 8-day global 500-meter surface reflectance product from 2002, excluding pixels that have low resolution due to broad-sensor zenith angles or dense clouds. The second procedure employs a wavelet transform to filter the EVI time series, found to be more effective than a Fourier transform, effectively eliminating noise. The third procedure determines various phenological stages, the overall growing period, and the dates for harvesting, heading, and planting. The highest EVI was used to determine the heading date. The planting date was identified by locating the inflection point or the minimal point that occurs within 60 days before the estimated heading date, as automatically determined by Sakamoto et al. (2005). Similarly, they determined the harvesting date as the inflection point occurring at least 30 days following the estimated heading date. Validation conducted at 30 Japanese test sites revealed that the RMSE values for phenological stages and growth period were approximately 12 days.

Sakamoto et al. (2005) applied the developed WFCP method in a follow-up investigation in the Red River Delta. Their findings indicated the presence of winter-spring rice-cropping zones in the eastern section and a double rice-cropping system in the delta. In another study conducted by the research group led by Sakamoto et al. (2005), the WFCP was utilized to analyze the spatial distribution of the heading stages and the rice-cropping system in the Mekong Delta. The study specifically examined how the system adapted to seasonal changes in water resources from 2002 to 2003. Modifications were made to the original WFCP, including the removal of the preprocessing step that accounted for the sensor zenith angle, which was necessary due to the significant rainfall during the rainy season. Additionally, the noise rejection threshold in the time series was adjusted in the second step of the method. This updated approach, referred to as the wavelet-based filter for assessing the spatial distribution in cropping systems (WFCS), was developed by an expert in crop management and remote sensing. The study observed an increase in triple-cropped rice cultivation in areas affected by salinity intrusion and flooding between 2002 and 2003. The results demonstrate the effective application of improved land management and advanced farming technologies in mitigating salinity and flood challenges.

Sakamoto et al. (2009) published two additional studies utilizing extended MODIS time series data from 2000 to 2006 and 2000 to 2008, respectively. A third method was introduced for estimating the start and end dates of flood inundation based on the EVI and land surface water index time series. This method called the wavelet-based filter for detecting spatiotemporal changes in flood inundation (WFFI), was added to the previously used WFCP and WFCS methods. By implementing this technique, researchers demonstrated a consistent increase in agricultural land and a transition from double to triple rice-cropping systems in the Mekong Delta. Infrastructure improvements, including water resource management and ring-dyke constructions, played a crucial role in enabling this shift (Šimon, 2004). While crop monitoring at the global level has been found to benefit from low-resolution remote sensing regional end users require a better resolution to make accurate decisions on, say, yield forecasting and estimating. The resolution of VEGETATION, MODIS, or AVHRR imagery is not precise enough to provide a detailed understanding of the field scale situation. Instead, these sensors are designed to provide a broad overview of the general conditions in the area. Nevertheless, the limited publications in the said research area suggest that there is still room for growth in utilizing low-resolution data (Puig et al., 2001). Certain studies aim to address the issue of low spatial resolution by utilizing more detailed data, such as linking data from VEGETATION or AVHHR to higher-resolution data. For instance, one study predicted the rice area in Hubei Province, China by using Landsat TM higher resolution data in conjunction with AVHRR data. Using a linear statistical model, data was linked to forecast the rice area for another year using AVHRR data. The usefulness of integrating low-resolution data appears doubtful because these methods still require higher-resolution data (Brill et al., 2024).

7.7.3. Hyperspectral Data-Based Remote Sensing of Rice Fields

With the advent of hyperspectral sensors, there have been significant advancements in accurately measuring spectral emittance or reflectance. These advancements have opened up new possibilities for quantitatively describing the agronomic parameters and gaining a deeper understanding of how various factors impact vegetation phenology (Shi et al., 2020). Several publications have utilized hyperspectral remote sensing to map rice crops. Over 70 publications were examined for this review, most of which dealt with field spectroradiometer applications. Field sensors and airborne/spaceborne sensors have a wide variety of discrete spectral bands in NIR and visible areas (400–2500 nm). One major benefit of using air-/spaceborne sensors is the consistent and extensive spatial coverage they provide, which is in contrast to the limited point data obtained from field spectroradiometers. However, spaceborne sensors require extensive preprocessing to correct for the geometric and atmospheric effects. This issue, particularly for spaceborne sensors, continues to be a significant topic in the literature. Spaceborne and airborne sensors have mainly been utilized for mapping rice and its cropping systems. On the other hand, field spectroradiometers are employed to study more intricate biophysical variables (Datt et al., 2003).

7.7.3.1. Fields of Application

1. **Managing Nutrients and Fertilizers:** In the literature, the analysis of yield components and grain yield is often linked with the examination of the amount and time of application of nitrogen fertilizers or the estimation of leaf nitrogen content. In this respect, a number of models have been used to forecast rice production based on markers of N status before heading. Previous research has highlighted the strong connection between the spectral transmittance of plant leaves and their chlorophyll or nitrogen content. In a later study, Inoue et al. (1998) discovered a significant linear relationship between the leaf nitrogen content and the spectral ratio R830 nm/R550 nm during the heading stage. Based on the paddy field being considered as a large leaf, this ratio appears to be a useful indicator for estimating the leaf nitrogen content. Inoue et al. (1998) utilized a blend of four spectral bands (R550 nm; R830 nm; R1650 nm; and R2200 nm) to estimate the total amount of leaf nitrogen (Apweiler et al., 2009).

   In a study conducted by Zhang et al. (2006), they examined the correlation between nitrogen supply and the spectral reflectance of the crop canopy. Their findings revealed notable variations in nitrogen concentrations among the leaves of the canopy during different stages of growth. They discovered a strong correlation between the NIR and visible spectrum ranges and the nitrogen status of rice. As a result, they utilized the RVI and the NDVI to forecast the nitrogen status of rice. A very similar work was carried out by Xue et al. (2004), who investigated the relationship between seasonal canopy reflectances, various band ratios, and NDVI and N accumulation and concentration in the rice leaves under various N treatments. The researchers discovered a strong linear relationship between the NIR and green band ratios (R810 nm/R560 nm) and N accumulation throughout the entire growth period, regardless of the level of treatment. Therefore, this band ratio was recommended in rice plants for the assessment of nitrogen status (Lucas et al., 2018).

In a study conducted by Lee et al. (2008), a straightforward spectral index (SI) was utilized to evaluate the nitrogen (N) status in rice plants. This index involved the initial derivative of the canopy reflectance at the value of 735 nm. Close linear relations were discovered in the panicle initiation stage (Cobo et al., 2011).

2. **Dealing with Insect Infestations and Managing Pests:** Multiple parameters such as crop growth stages, and N treatment; along with the presence of pest and weed infestations, can cause spectral interferences. These interferences make it challenging to accurately distinguish between different species. *Nilaparavata lugens*, the brown planthopper, is widely recognized as a significant insect pest in numerous rice-producing areas. It causes extensive harm to rice crops annually. Controlling these insects can be quite challenging for rice farmers. The association between spectral traits and signs of infestations of the Brown Planthopper and another extensively dispersed pest, *Cnaphalocrocis medinalis*, the rice leaffolder, was examined by Yang & Cheng (2001) and later by Chen et al. (2007). Spectral reflectance and infestation scale were found to be linearly correlated for both band ratios and single bands. According to Chen et al. (2007), canopies infested with brown planthopper showed the greatest band ratio determination coefficients R_{NIR}/R_{red} ($R^2 = 0.922$) and the highest correlation coefficients at 426 nm ($r = 0.878$). Canopies infested with leaffolders displayed a range of negative correlations, which varied based on the rice plants' growth stage. For site-specific pest control, a deeper comprehension of the field conditions and interactions between pests like fungi, insects, and plants is possible through the study of hyperspectral data. In addition, it can enhance pesticide application timings and the estimation of the required quantity (Sandier, 1998).

3. **Managing Irrigation:** Droughts pose a significant challenge in paddy rice cultivations, in rainfed conditions particularly. Utilizing

remote sensing technology can enhance the monitoring of water stress and drought damage, leading to more effective management of water applications. To this end, Shibayama et al. (1993) examined the physiological changes in the rice plants brought on by water shortages as well as the spectral assessment of variations in soil water content. Even in the presence of a dense canopy cover, they discovered that the mid-infrared and NIR bands (1190–1320 nm and around 1600 nm) remained extremely sensitive to soil background and water surfaces. Nevertheless, the sensitivity observed is a result of the water background and not a result of any physiological shifts in the rice canopy. The first derivative of the canopy reflection spectrum was used to find out the water status of the canopies. This gets rid of any background impacts. According to their research, it was discovered that the derivative at precisely 960 nm was the most effective for detecting water-stressed canopies. This method was able to identify the stress before any visible symptoms appeared, even before the NDVI method (Coste et al., 1994).

In a separate study, Song et al. (2011) examined the spectral reflectance from paddy rice under water stress conditions. According to their study, bands at certain wavelengths (1158 nm; 1378 nm; and 1965 nm) were identified as highly valuable for detecting stress conditions across various irrigation levels. However, because leaf sample water was lost during the measurement experiment, the authors claim that the three-band regression's accuracy is fairly low ($R^2 = 0.75$).

4. **Dealing with Heavy Metal Pollution:** The rapid pace of industrial development in emerging countries like China has led to numerous environmental challenges. Contamination of heavy metals in soils is a significant issue that can have serious health implications when it comes to agricultural crops. In their study, various indicators based on the hyperspectral data were examined to identify rice stress caused by heavy metal pollution. They concluded that the red-edge position (REP) is a highly sensitive indicator for this particular purpose. Depending on the level of pollution, they observed shifts of 3 and 6 nm toward the shorter wavelength (Machuca & Phillips, 1983).

7.7.3.2. Methods

The hyperspectral field research techniques need to be adapted for use on aircraft or spacecraft in order to cover greater ground. Despite its immense potential in analyzing paddy rice fields, there is a lack of comprehensive literature on hyperspectral remote sensing from spaceborne and airborne hyperspectral remote sensing. This is likely because of the expensive data costs and the limited availability of spaceborne hyperspectral sensors (Dawson et al., 1999). In a study conducted by LaCapra et al. (1996), the foliar chemistry of multiple rice fields in California, USA was analyzed using the airborne AVIRIS hyperspectral sensor. To forecast the concentrations of lignin and nitrogen from the spectrum of reflectance modified as the first difference of absorbance, they used a stepwise multiple linear regression. They were unable to find calibration e□uations that were statistically significant for the concentration of lignin in rice plants. Given the wide range of chemical varieties and relatively low concentrations of lignins, it is not surprising that these results were consistent with previous findings. However, it was demonstrated that airborne hyperspectral data has the potential to anticipate the concentration of nitrogen in the paddy rice fields. The study found that a general calibration equation could not be derived due to the use of distinct sets of wavelengths to represent various subsets of the data in their multilevel relational data (MLR) tests. In their study, Ryu et al. (2011) utilized an airborne hyperspectral sensor called the AISA Eagle to analyze the concentration of nitrogen in the rice plants at a research site in Japan. They specifically emphasized the nitrogen concentration during the heading stage, as it is believed to have a significant impact on the □uantity and □uality of rice grains. They conducted experiments using various periods, ranging from one to three years, utilizing partial least square regression or PLSR and MLR. The findings suggest that PLSR models exhibit a higher level of sensitivity towards nitrogen concentration during the heading stage, particularly when considering data from multiple years (Bouvet et al., 2009).

7.8. SUMMARY

This chapter provided an introduction to mapping rice-cropped areas using remote sensing techniques, along with a comprehensive review of relevant studies. Depending on the study's application field, different sensors are employed for this mapping. One of the most common uses is to accurately identify and map rice paddies to determine their total coverage. For effective monitoring and trend analysis over multiple growing seasons, it is essential to implement a change detection method using classified images. Additionally, remote sensing is used to distinguish between rainfed and irrigated rice, as well as to identify various cultural practices. Furthermore, specialized applications of satellite and airborne remote sensing include identifying breeding habitats of malaria vectors, detecting residue burning and its associated greenhouse gas (GHG) emissions, and estimating and modeling rice yields. Field spectroradiometers are commonly used for applications requiring detailed spectral information, such as fertilizer or nutrient management, or for detecting water stress.

REFERENCES

1. Amarasinghe, U. S., & Weerakoon, D. E. M. (2009). Present status and future strategies for the management of reservoir fisheries in Sri Lanka. *Status of Reservoir Fisheries in Five Asian Countries,* (2), 55–69.
2. Apweiler, R., Aslanidis, C., Deufel, T., Gerstner, A., Hansen, J., Hochstrasser, C, & Schmitz, G. (2009). Approaching clinical proteomics: Current state and future fields of application in fluid proteomics. *Clinical Chemistry and Laboratory Medicine, 47*(6), 724–744.
3. Aschbacher, J., Pongsrihadulchai, A., Karnchanasutham, S., Rodprom, C., Paudyal, D. R., & Le Toan, T. (1995, July). Assessment of ERS-1 SAR data for rice crop mapping and monitoring. In *1995 International Geoscience and Remote Sensing Symposium (IGARSS)* (Vol. 3, pp. 2183–2185). Quantitative Remote Sensing for Science and Applications.
4. Badarinath, K. V. S., Latha, K. M., Chand, T. K., & Gupta, P. K. (2009). Impact of biomass burning on aerosol properties over tropical wet evergreen forests of Arunachal Pradesh, India. *Atmospheric Research, 91*(1), 87–93.
5. Bailey, K. W., Jones, C. M., & Heinrichs, A. J. (2005). Economic returns to Holstein and Jersey herds under multiple component pricing. *Journal of Dairy Science, 88*(6), 2269–2280.
6. Berth, P., & Jeschke, P. (1989). Consumption and fields of application of LAS. *Tenside Surfactants Detergents, 26*(2), 75–79.
7. Bouman, B. (2009). How much water does rice use? *Management, 69*(2), 115–133.
8. Bouvet, A., Le Toan, T., & Lam-Dao, N. (2009). Monitoring of the rice cropping system in the Mekong Delta using ENVISAT/ASAR dual polarization data. *IEEE Transactions on Geoscience and Remote Sensing, 47*(2), 517–526.
9. Bresler, E., & Dagan, G. (1983). Unsaturated flow in spatially variable fields: 2. Application of water flow models to various fields. *Water Resources Research, 19*(2), 421–428.
10. Brill, M., Freeman, R., Janson, S., & Lackner, M. (2024). Phragmén's voting methods and justified representation. *Mathematical Programming, 203*(1), 47–76.
11. Cao, M., Gregson, K., Marshall, S., Dent, J. B., & Heal, O. W. (1996). Global methane emissions from rice paddies. *Chemosphere, 33*(5), 879–897.

12. Che Salmah, M. R., Siregar, A. Z., Abu Hassan, A., & Nasution, Z. J. T. E. (2017). Dynamics of a□uatic organisms in a rice field ecosystem: Effects of seasons and cultivation phases on abundance and predator-prey interactions. *Tropical Ecology, 58*(1), 177–191.
13. Chen, C., & Mcnairn, H. (2006). A neural network-integrated approach for rice crop monitoring. *International Journal of Remote Sensing, 27*(7), 1367–1393.
14. Chen, H. Y., Yu, S. L., Chen, C. H., Chang, G. C., Chen, C. Y., Yu, A., & Yang, P. C. (2007). A five-gene signature and clinical outcome in non–non-small-cell lung cancer. *New England Journal of Medicine, 356*(1), 11–20.
15. Cobo, M. J., López-Herrera, A. G., Herrera-Viedma, E., & Herrera, F. (2011). An approach for detecting, quantifying, and visualizing the evolution of a research field: A practical application to the Fuzzy Sets Theory field. *Journal of Informetrics, 5*(1), 146–166.
16. Coste, J., Frerot, E., & Jouin, P. (1994). Coupling N-methylated amino acids using PyBroP and PyCloP halogenophosphonium salts: Mechanism and fields of application. *The Journal of Organic Chemistry, 59*(9), 2437–2446.
17. Darvishsefat, A. A., Abbasi, M., & Schaepman, M. E. (2011). Evaluation of spectral reflectance of seven Iranian rice varieties canopies. *Journal of Agricultural Science and Technology, 13,* 1091–1104.
18. Datt, B., McVicar, T. R., Van Niel, T. G., Jupp, D. L., & Pearlman, J. S. (2003). Preprocessing EO-1 Hyperion hyperspectral data to support the application of agricultural indexes. *IEEE Transactions on Geoscience and Remote Sensing, 41*(6), 1246–1259.
19. Dawson, T. P., Curran, P. J., North, P. R. J., & Plummer, S. E. (1999). The propagation of foliar biochemical absorption features in forest canopy reflectance: A theoretical analysis. *Remote Sensing of Environment, 67*(2), 147–159.
20. De Datta, S. K. (1981). *Principles and practices of rice production.* Manila: The International Rice Research Institute. *(Vol. 6, pp. 100–400).*
21. Diuk - Wasser, M. A., Dolo, G., Bagayoko, M., Sogoba, N., Toure, M. B., Mogham, M., & Taylor, C. E. (2006). Patterns of irrigated rice growth and malaria vector breeding in Mali using multi - temporal ERS - 2 synthetic aperture radar. *International Journal of Remote Sensing, 27*(3), 535–548.
22. Dong, Y., Sun, G., & Pang, Y. (2006). Monitoring of rice crop using ENVISAT ASAR data. *Science in China Series D: Earth Sciences, 49,* 755–763.

23. Duan, B., Liu, Y., Gong, Y., Peng, Y., Wu, X., Zhu, R., & Fang, S. (2019). Remote estimation of rice LAI based on Fourier spectrum texture from UAV image. *Plant Methods, 15,* 1–12.

24. Fairhurst, T., & Dobermann, A. (2002). Rice in the global food supply. *World, 5*(7,502), 454–349.

25. Faisal, K., AlAhmad, M., & Shaker, A. (2012). Remote sensing techniques as a tool for environmental monitoring. *The International Archives of the Photogrammetry, Remote Sensing and Spatial Information Sciences, 39,* 513–518.

26. Fang, J., Chen, A., Peng, C., Zhao, S., & Ci, L. (2001). Changes in forest biomass carbon storage in China between 1949 and 1998. *Science, 292*(5525), 2320–2322.

27. Frolking, S., Qiu, J., Boles, S., Xiao, X., Liu, J., Huang, Y., & Qin, X. (2002). Combining remote sensing and ground census data to develop new maps of the distribution of rice agriculture in China. *Global Biogeochemical Cycles, 16*(4), 38–1.

28. Gandharum, L., Mulyani, M. E., Hartono, D. M., Karsidi, A., & Ahmad, M. (2021). Remote sensing versus the area sampling frame method in paddy rice acreage estimation in Indramayu regency, West Java province, Indonesia. *International Journal of Remote Sensing, 42*(5), 1738–1767.

29. Gumma, M. K., Gauchan, D., Nelson, A., Pandey, S., & Rala, A. (2011). Temporal changes in the rice-growing area and their impact on livelihood over a decade: A case study of Nepal. *Agriculture, Ecosystems & Environment, 142*(3-4), 382–392.

30. Hedley, J. D., Roelfsema, C. M., Chollett, I., Harborne, A. R., Heron, S. F., J. Weeks, S., & Mumby, P. J. (2016). Remote sensing of coral reefs for monitoring and management: A review. *Remote Sensing, 8*(2), 118.

31. Heong, K. L. (2008, November). Biodiversity, ecosystem services, and pest management. In *Second International Plantation Industry Conference and Exhibition (IPiCEX),* Shah Alam, November (pp. 18–21).

32. Huete, A., Didan, K., Miura, T., Rodriguez, E. P., Gao, X., & Ferreira, L. G. (2002). Overview of the radiometric and biophysical performance of the MODIS vegetation indices. *Remote Sensing of Environment, 83*(1–2), 195–213.

33. Inoue, A., Seidel, M. G., Wu, W., Kamizono, S., Ferrando, A. A., Onson, R. T., & Look, A. T. (2002). Slug, a highly conserved zinc finger transcriptional repressor, protects hematopoietic progenitor cells from radiation-induced apoptosis in vivo. *Cancer Cell, 2*(4), 279–288.

34. Inoue, Y., Moran, M. S., & Horie, T. (1998). Analysis of spectral measurements in paddy field for predicting rice growth and yield based on a simple crop simulation model. *Plant Production Science, 1*(4), 269–279.
35. Ishiguro, K., Shiratsuchi, A., Sato, S., Omori, A., Arioka, C, Kobayashi, S., & Imahori, K. (1993). Glycogen synthase kinase 3β is identical to tau protein kinase I generating several epitopes of paired helical filaments. *FEBS Letters, 325*(3), 167–172.
36. Islam, M. R. (2021). Irrigated rice area mapping over Bangladesh with remotely sensed data from 2001 to 2018. *Agrotechnology, 10,* 213.
37. Jiang, X., Tang, L., Liu, X., Huang, F., Cao, W., & Zhu, Y. (2011). Spatial and temporal characteristics of rice production climatic resources in main growing regions of China. *Transactions of the Chinese Society of Agricultural Engineering, 27*(7), 238–245.
38. Kim, K. S., Zhao, Y., Jang, H., Lee, S. Y., K. J. M., Kim, K. S., & Hong, B. H. (2009). Large reebankpattern growth of graphene films for stretchable transparent electrodes. *Nature, 457*(7230), 706–710.
39. Krupa, M., Tate, K. W., van Kessel, C., Sarwar, N., & Linquist, B. A. (2011). Water quality in rice-growing watersheds in a Mediterranean climate. *Agriculture, Ecosystems & Environment, 144*(1), 290–301.
40. Kuenzer, C., & Knauer, K. (2013). Remote sensing of rice crop areas. *International Journal of Remote Sensing, 34*(6), 2101–2139.
41. LaCapra, D. (1999). Trauma, absence, loss. *Critical Inquiry, 25*(4), 696–727.
42. Lampayan, R. M., Faronilo, J. E., Tuong, T. P., Espiritu, A. J., De D., J. L., Bayot, R. S., & Hosen, Y. (2015). Effects of seedbed management and delayed transplanting of rice seedlings on crop performance, grain yield, and water productivity. *Field Crops Research, 183,* 303–314.
43. Le Toan, T., Ribbes, F., Wang, L. F., Floury, N., Ding, K. H., Kong, J. A., & Kurosu, T. (1997). Rice crop mapping and monitoring using ERS-1 data based on experiment and modeling results. *IEEE Transactions on Geoscience and Remote Sensing, 35*(1), 41–56.
44. Lee, J., Lee, S., & Young, J. P. W. (2008). Improved PCR primers for the detection and identification of arbuscular mycorrhizal fungi. *FEMS Microbiology Ecology, 65*(2), 339–349.
45. Lucas-González, R., Viuda-Martos, M., Pérez-Alvarez, J. A., & Fernández-López, J. (2018). In vitro digestion models suitable for foods: Opportunities for new fields of application and challenges. *Food Research International, 107,* 423–436.

46. Machuca, R., & Phillips, K. (1983). Applications of vector fields to image processing. *IEEE Transactions on Pattern Analysis and Machine Intelligence, (3),* 316–329.
47. Mansaray, L. R., Huang, W., Zhang, D., Huang, J., & Li, J. (2017). Mapping rice fields in urban Shanghai, southeast China, using Sentinel-1A and Landsat 8 datasets. *Remote Sensing, 9*(3), 257.
48. McCloy, K. R., Smith, F. R., & Robinson, M. R. (1987). Monitoring rice areas using Landsat MSS data. *International Journal of Remote Sensing, 8*(5), 741–749.
49. Mosleh, M. K., Hassan, Q. K., & Chowdhury, E. H. (2015). Application of remote sensors in mapping rice area and forecasting its production: A review. *Sensors, 15*(1), 769–791.
50. Motohka, T., Nasahara, K. N., Miyata, A., Mano, M., & Tsuchida, S. (2009). Evaluation of optical satellite remote sensing for rice paddy phenology in monsoon Asia using a continuous in situ dataset. *International Journal of Remote Sensing, 30*(17), 4343–4357.
51. Oguro, H., Yuan, J., Ichikawa, H., Ikawa, T., Mazaki, S., Kawamoto, H., & Iwama, A. (2010). Poised lineage specification in multipotential hematopoietic stem and progenitor cells by the polycomb protein Bmi1. *Cell Stem Cell, 6*(3), 279–286.
52. Oyoshi, K., Tomiyama, N., Okumura, T., Sobue, S., & Sato, J. (2016). Mapping rice-planted areas using time-series synthetic aperture radar data for the Asia-RiCE activity. *Paddy and Water Environment, 14,* 463–472.
53. Panigrahy, R. K., Ray, S. S., & Panigrahy, S. (2009). Study on the utility of IRS-P6 AWIFS SWIR band for crop discrimination and classification. *Journal of the Indian Society of Remote Sensing, 37,* 325–333.
54. Panigrahy, S., & Sharma, S. A. (1997). Mapping of crop rotation using multidate Indian Remote Sensing Satellite digital data. *ISPRS Journal of Photogrammetry and Remote Sensing, 52*(2), 85–91.
55. Panigrahy, S., Kumar, T., & Manjunath, K. R. (2012). Hyperspectral leaf signature as an added dimension for species discrimination: A case study of four tropical mangroves. *Wetlands Ecology and Management, 20,* 101–110.
56. Patel, J., Zeckhauser, R., & Hendricks, D. (1991). The rationality struggle: Illustrations from financial markets. *The American Economic Review, 81*(2), 232–236.
57. Puig, O., Caspary, F., Rigaut, G., Rutz, B., Bouet, E., Bragado-Nilsson, E., & Séraphin, B. (2001). The tandem affinity purification (TAP) method: A general procedure of protein complex purification. *Methods, 24*(3), 218–229.

58. Ribbes, F. (1999). Rice field mapping and monitoring with RADARSAT data. *International Journal of Remote Sensing, 20*(4), 745–765.

59. Ryu, C., Suguri, M., & Umeda, M. (2011). Multivariate analysis of nitrogen content for rice at the heading stage using reflectance of airborne hyperspectral remote sensing. *Field Crops Research, 122*(3), 214–224.

60. Sakamoto, J., van Heijst, J., Lukin, O., & Schlüter, A. D. (2009). Two - dimensional polymers: Just a dream of synthetic chemists? *Angewandte Chemie International Edition, 48*(6), 1030–1069.

61. Sakamoto, T., Yokozawa, M., Toritani, H., Shibayama, M., Ishitsuka, N., & Ohno, H. (2005). A crop phenology detection method using time-series MODIS data. *Remote Sensing of Environment, 96*(3-4), 366–374.

62. Sandier, E. (1998). Lower bounds for the energy of unit vector fields and applications. *Journal of Functional Analysis, 152*(2), 379–403.

63. Shen, J., Chanda, A., D'netto, B., & Monga, M. (2009). Managing diversity through human resource management: An international perspective and conceptual framework. *The International Journal of Human Resource Management, 20*(2), 235–251.

64. Shi, Y., Lou, Y., Zhang, Z., Ma, L., & Ojara, M. A. (2020). Estimation of methane emissions based on crop yield and remote sensing data in a paddy field. *Greenhouse Gases: Science and Technology, 10*(1), 196–207.

65. Shibayama, M., Takahashi, W., Morinaga, S., & Akiyama, T. (1993). Canopy water deficit detection in paddy rice using a high-resolution field spectroradiometer. *Remote Sensing of Environment, 45*(2), 117–126.

66. Šimon, P. (2004). Isoconversional methods. *Journal of Thermal Analysis and Calorimetry, 76,* 123–132.

67. Singh, R. P., Oza, S. R., & Pandya, M. R. (2006). Observing long-term changes in rice phenology using NOAA–AVHRR and DMSP–SSM/I satellite sensor measurements in Punjab, India. *Current Science, 1217–1221.*

68. Smith, K. C. A., & Oatley, C. W. (1955). The scanning electron microscope and its fields of application. *British Journal of Applied Physics, 6*(11), 391.

69. Song, Z., Storesletten, K., & Zilibotti, F. (2011). Growing like China. *American Economic Review, 101*(1), 196–233.

70. Tan, S., Chen, B., Zg, H., Xu, F., Xu, H., Wei, J., & Tang, J. (2019). Effects of cultivation techniques on CH4 emissions, net ecosystem production, and rice yield in a paddy ecosystem. *Atmospheric Pollution Research, 10*(1), 274–282.

71. Tennakoon, S. B., Murty, V. V. N., & Eiumnoh, A. (1992). Estimation of cropped area and grain yield of rice using remote sensing data. *International Journal of Remote Sensing, 13*(3), 427–439.

72. Tu, Y. (2008). The nonequilibrium mechanism for ultrasensitivity in a biological switch: Sensing by Maxwell's demons. *Proceedings of the National Academy of Sciences, 105*(33), 11737–11741.

73. Tuong, T. P., & Bouman, B. A. (2003). Rice production in water-scarce environments. *Water Productivity in Agriculture: Limits and Opportunities for Improvement, 1,* 13–42.

74. Turner, M. D., & Congalton, R. G. (1998). Classification of multi-temporal SPOT-XS satellite data for mapping rice fields on a West African floodplain. *International Journal of Remote Sensing, 19*(1), 21–41.

75. Usali, N., & Ismail, M. H. (2010). Use of remote sensing and GIS in monitoring water quality. *Journal of Sustainable Development, 3*(3), 228.

76. Van Niel, T. G., & McVicar, T. R. (2004). Current and potential uses of optical remote sensing in rice-based irrigation systems: A review. *Australian Journal of Agricultural Research, 55*(2), 155–185.

77. Verbano, C., & Venturini, K. (2011). Development paths of risk management: Approaches, methods, and fields of application. *Journal of Risk Research, 14*(5), 519–550.

78. Xiao, X., Shao, S. H., Huang, Z. D., & Chou, K. C. (2006). Using pseudo amino acid composition to predict protein structural classes: Approached with complexity measure factor. *Journal of Computational Chemistry, 27*(4), 478–482.

79. Xue, N., Xia, F., Chiou, reeban, reebank, M. (2005). The Penn Chinese Treebank: Phrase structure annotation of a large corpus. *Natural Language Engineering, 11*(2), 207–238.

80. Xue, Y., Chang, L., Kjaer, S. B., Bordonau, J., & Shimizu, T. (2004). Topologies of single-phase inverters for small distributed power generators: An overview. *IEEE Transactions on Power Electronics, 19*(5), 1305–1314.

81. Yang, F. Y., & Cheng, C. H. (2001). Unusual diboration of allenes catalyzed by palladium complexes and organic iodides: A new efficient route to biboronic compounds. *Journal of the American Chemical Society, 123*(4), 761–762.

82. Ze, H. E., & Shihua, L. I. (2023). Research progress on radar remote sensing for rice growth monitoring. *National Remote Sensing Bulletin, 27*(10), 2363–2382.

83. Zhang, X., Yazaki, J., Sundaresan, A., Kus, S., Chan, S. W. L., Chen, H., & Ecker, J. R. (2006). Genome-wide high-resolution mapping and functional analysis of DNA methylation in *Arabidopsis*. *Cell, 126*(6), 1189–1201.

Chapter 8

REMOTE SENSING AS A MANAGEMENT AND MONITORING TOOL

Contents

8.1. INTRODUCTION

Pests and plant diseases pose serious risks to forestry and global agriculture. Understanding the severity, extent, and location of pests and diseases is crucial for effectively directing the procedures for the protection of plants. Considering the drawbacks of traditional field scouting for pests and plant diseases, such as being labor-intensive, inefficient, and subjective, the applications of remote sensing (RS) methods could significantly improve the monitoring of these issues on a larger scale (Mahlein, 2015).

Remote sensing (RS) of pests and plant diseases can be considered a "radiodiagnosis" method for plants, offering efficient, spatially continuous, and non-contact monitoring. Early studies and applications in this field date back to the 1980s. According to Riley (1989), it is possible to identify areas affected by pests and plant diseases through visual interpretation of aerial or satellite images. Nilsson (1995) demonstrated multiple uses of image analysis and RS in the field of plant pathology. As sensing technologies and computer science have advanced rapidly, various types of RS data are now used to detect pests and diseases. Over the past few decades, extensive research has focused on sensing systems, algorithms, and feature extraction at various scales, revealing exciting opportunities for the remote monitoring of pests and plant diseases.

Through this method, the relationship between the theories in plant pathology and remote sensing methods has been enhanced, leading to a deeper comprehension of the agricultural system in various positions. For instance, previous research has not only concentrated on identifying the presence of a particular disease or pest but has also aimed to differentiate between various pests and diseases simultaneously, evaluate the severity of their infections, and create maps of their distributions at both regional and plot scales. Techniques identified for modeling and feature extraction extend beyond traditional statistical discriminant and spectral features methods. They incorporate innovative intelligent ML algorithms and signal processing techniques. There are certain instances where the monitoring and the detection of pests/diseases have proven to be highly valuable in various practical scenarios. For example, they can assist in the field to accurately target sprays for pest and disease management in the field, aid in the efficient phenotyping of plant characteristics, or contribute to evaluating losses for agricultural insurance purposes (Zhang et al., 2015).

8.2. PRINCIPLES OF PEST AND DISEASE MONITORING FOR PLANTS

When studying the interactions between pathogens and hosts, various crop pests and diseases can lead to a range of symptoms and damage to plants. These physical changes serve as the foundation for monitoring them through remote sensing. It is important to keep in mind that all plant pests and diseases can be detected using remote sensing, as a few of them do not have identifiable characteristics. On the flip side, certain root diseases along with soil-based diseases can also be detected, 'which have systemic effects on the plant's physiology. Thus, a crucial factor in the monitoring and detection of plant pests/diseases through RS is the existence of a distinct reaction that can be identified by a specific sensor system or a sensor (MacDougall et al., 2022).

Four categories of damages related to remote sensing have been identified among the physiological changes and symptoms of diseases in plants induced by pests or diseases (Figure 8.1):

1. **Loss of LAI and Reduction of Biomass:** This kind of destruction typically happens during pest infestations. Pests such as armyworms can devour various parts of plants, including leaves and stalks, resulting in a substantial reduction in the biomass and leaf area. Nevertheless, considering the lack of spectral specificity in this destruction, monitoring it always presents the challenge of high uncertainty.
2. **Infection-Related Lesion or Pustule:** The most prevalent symptoms are pustules or lesions, which are necrotic or sori tissues brought on by pests and diseases. They can exhibit a range of colors and shapes, which can differ depending on the specific pests or diseases. It is important to consider that the abundance as well as distribution of these pustules and lesions inside the canopy, that is, focused on the bottom of the canopy or distributed uniformly, can greatly affect how easily they can be detected (Barzman et al., 2015).
3. **Disruption of the Pigment Systems:** Often, the attacks from pests or infection of diseases can result in a breakdown of chloroplasts or other organelles. This can lead to changes in the levels of pigments such as anthocyanins, carotenoids (Car), or chlorophyll (Chl). Typically, hyperspectral RS observations may be necessary to recognize the response of this kind (Cook et al., 1996).
4. **Wilting of Plant Leaves:** Dehydration-induced loss of rigidity is not a commonly observed symptom of plant pests and diseases, often leading to confusion with drought stress. However, certain pests, such as aphids or beetles, can cause plants to wilt due to their sucking and piercing activities. In some cases of severe infection, the plant's vascular system may become damaged, leading to a blockage of water flow and subse□uent dehydration of the entire plant (Prabhakar et al., 2022).

It is important to highlight that the presence of plant pests or diseases can lead to various forms of damage, as discussed earlier. Remote sensing should be capable of capturing the buildup of symptoms. Since pest attacks or disease infections typically follow a temporal pattern, various symptoms may overlap or interact with one another, appearing with different levels of severity at different stages. For example, the early stages of winter wheat yellow rust may result in leaf lesions and a reduction in chlorophyll concentration. In the absence of adequate control, the impaired photosynthetic system can lead to reduced biomass and hinder the plant's water metabolism (Carvajal et al., 2019). As the infection progresses, plants will exhibit wilting symptoms in the later stages.

Figure 8.1. *Examples of plant damage from pests and diseases detected via remote sensing.*

Note: Sample images of some pests – (A) rice leaf roller; (B) rice leaf caterpillar; (C) paddy stem maggot; (D) rice water weevil; (E) rice leafhopper; (F) grain spreader thrips; (G) yellow cutworm; (H) red spider; (I) corn borer; (J) wheat blossom midge; (K) Penthaleus major; and (L) long-legged spider mite.

Understanding the temporal pattern of symptom progression poses challenges for monitoring and detecting pests and diseases. Therefore, it is important to identify specific remote sensing features at various stages of pest attack or disease. On the positive side, the occurrence of symptoms at different times can provide valuable information. By identifying these temporal patterns and using them as unique signatures for pests or diseases, uncertainty in monitoring efforts can be minimized (Ratnadass et al., 2012).

8.3. THE PEST AND PLANT DISEASE MONITORING REMOTE SENSING SYSTEMS

Several RS systems could be used to monitor and detect pest/plant disease. This RS equipment is capable of acquiring data from microwave to gamma rays and can function using active and passive radiations.

To effectively monitor and detect pests/plant diseases, various remote sensing systems have been utilized to capture the infection symptoms including scabs and pustules; the physiological responses including changes in water and pigment content, and the structural changes including the landscapes and canopy structures caused by pests/plant diseases (Hahn, 2009). When it comes to monitoring pests/plant diseases, the sensing systems may be broadly divided into the following three groups based on technical maturity and sensing principles:

- Visible and near-infrared or VIS-SWIR spectral methods;
- Fluorescence and thermal methods;
- SAR and lidar systems (Table 8.1) (Terentev et al., 2022).

When it comes to wavelengths shorter than visible bands, such as gamma rays, X-rays, and UV rays, the sensors typically operate in an active mode, which can limit their portability.

These sensors are primarily used for indoor testing of pests and plant diseases, especially in fruits and vegetables. It is important to note that the categorization of sensing systems mentioned does not strictly adhere to sensing modes or wavelength ranges.

However, it does emphasize the crucial role of sensing systems in monitoring pests and plant diseases (Gogoi et al., 2018). Data characteristics and scales of different data sources used for remote sensing monitoring of rice diseases and pests are detailed in Table 8.2.

Table 8.1. *Pest and Plant Disease Monitoring Using Remote Sensing Systems*

Data Source	Data Characteristics	Data Acquisition Equipment	Monitoring Scale
Non-imaging hyperspectral	High spectral resolution and rich bands	ASD (analytical spectral devices) FieldSpec spectrometer	Leaf scale, canopy scale, field scale
Image hyperspectral	The high spectral resolution, rich bands, and image information	Scanning imaging spectrometer, airborne hyperspectral imager	Leaf scale, canopy scale, field scale, regional scale
Multispectral	Large monitoring range, low cost	Multispectral imaging camera (MS-4100), Satellite imagery (Landsat, TM, Sentinel-2A/B, Quickbird, WorldView-2, HJ-CCD, IKONOS), UAV (unmanned aerial vehicle) multispectral image	Field-scale, regional scale, global scale
Imaging	Rich disease and pest symptom information	Camera, Imaging spectrometer (Headwall), UAV hyperspectral/multispectral sensor	Leaf scale, canopy scale, field scale
Fluorescence	Sensitive pointers for photosynthetic functions	Fluorescence spectrum (PAM-2100), IMAGING-PAM	Leaf scale, canopy scale
Thermal infrared	Quick collection, non-destructive	Thermal infrared imager (FLIR), thermal infrared satellite image (TM, ASTER, HJ-IRS)	Field-scale, regional scale

Source: Zheng, Q., Huang, W., Xia, Q., Dong, Y., Ye, H., Jiang, H., Chen, S., & Huang, S. (2023). Remote sensing monitoring of rice diseases and pests from different data sources: A review. Agronomy, 13(7), 1851. https://doi.org/10.3390/agronomy13071851.

8.3.1. The VIS-SWIR System

Given that solar energy is most prominent in the visible alongside near-infrared, or VIS-NIR bands, sensors that utilize such bands are typically able to capture data of higher quality with a favorable signal-to-noise ratio. The VIS-NIR sensor being a well-established and cost-effective technology can be easily found on various remote sensing platforms such as satellite-, aerial-, or ground-based platforms. This makes them highly suitable for monitoring pests and plant diseases (Frank et al., 1995). Different physiological changes and signs of pests and diseases including scabs, cellular damage, and pigment degradation exhibit distinct responses in spectrum reflectance in the spectral domain of VIS-NIR. Furthermore, certain bands within the SWIR region, with wavelengths ranging from 1000 to 2500 nm, can effectively measure water content in soil or plants. These bands serve as valuable complements to traditional VIS-NIR sensors, which operate within the wavelength range of 400 to 1000 nm. By utilizing SWIR bands, researchers can gain insights into habitat-related characteristics of pests and plant diseases, as demonstrated in studies from (Xu et al., 2007).

With the inclusion of SWIR channels in advanced RS satellite sensors like Sentinel-2; HJ-CCD; Landsat 8-TM, and so forth, there is a growing anticipation for further research in this area. Both multi- and hyper-spectral sensors are extensively utilized to monitor pests/plant diseases since they are crucial branches of VIS-NIR sensors. The hyperspectral sensor with its higher spectral resolution enables the detection of subtle spectral changes, which is crucial for early-stage monitoring of pests and plant diseases, even in complex scenarios. Several studies have shown that multi-spectral remote sensing data, especially with a higher temporal and spatial resolution, can be efficient in certain monitoring situations. Furthermore, when considering hyperspectral versus multi-spectral data, it is worth noting that multi-spectral data currently offers several advantages. These include lower cost; finer spatial resolution; and wider availability. As a result, multi-spectral data can be a suitable choice for certain operational monitoring applications (Vernon & Sandini, 1988).

8.3.2. Fluorescence and Thermal Systems

With the use of thermal infrared and fluorescence remote sensing systems, it becomes possible to closely monitor the photosynthetic and respirational processes of plants. This advanced technology enables an early detection of pests/plant diseases, even before any symptoms become apparent. The most promising technique for obtaining the plant's fluorescence plants is laser-induced fluorescence (LIF). Certain commercial fluorometers, like the portable chlorophyll fluorometer of the PAM series (IMAGINE-PAM, PAM2500, and PAM2100, etc.), use the induced LED light sources at the fixed wavelengths (655 and 735 nm) along saturation pulse techniques to record various kinetics parameters and curves and to generate some significant fluorescence indicators, such as the operational efficiency of PSII (\\PSII); non-photochemical fluorescence quenching or NPQ; the maximal quantum yield (Fv/Fm), and so forth. Furthermore, when a laser-emitting light source (such as a Helium-Neon laser) is combined with an analytical spectrometer (like the FieldSpec TM UV/VNIR), it becomes possible to measure continuous fluorescence spectra. This enables the fluorescence features extraction at various wavelengths. Römer et al. (2011) used a robust polynomial fitting approach with fluorescence spectra covering 370–800 nm to obtain a highly accurate, that is, 93% pre-symptomatic disease identification in the wheat leaf rust (Barilero et al., 2009).

Furthermore, the presence of pests and diseases can alter the ETa and respiration rates in the plants, resulting in detectable changes in the levels of thermal emission. Researchers have discovered that thermal imagery may effectively monitor temperature irregularities in grapevines, allowing for early detection of fungal pathogens like *Plasmopara viticola*. By analyzing thermal images, Falkenberg et al. (2007) were able to distinguish between abiotic, for example, drought stresses, and biotic, for example, root stresses in cotton. Several studies have also confirmed the effectiveness of fluorescence as well as thermal techniques for early detection of pests and plant diseases (Yamauchi et al., 2006).

Nevertheless, there are certain drawbacks when it comes to utilizing fluorescence and thermal RS systems to monitor pests/plant diseases. First off, the productivity and viability of this method are limited at a large scale since it necessitates dark adaptation to the plant to measure several significant fluorescence characteristics like ΦPSII; NPQ; and Fv/Fm. In addition, considering the relatively weak fluorescence signals and their potential confusion with natural light, the majority of existing fluorometers have been specifically designed for measuring at the canopy or leaf levels. As a result, their usage in large areas has been limited. One effective approach to enhance the monitoring of pest and plant diseases is combining fluorescence and thermal techniques with other remote sensing systems, like hyperspectral systems. Recent studies have made progress in extracting fluorescence signals from the canopy spectra, which presents an exciting opportunity to use sensor fusion approaches for mapping plant diseases and pests. These approaches can include hyperspectral/multi-spectral, thermal, and fluorescence maps. According to research conducted by Bauriegel and Herppich in 2014, the monitoring system is anticipated to benefit from the high specificity and early-stage sensitivity offered by both systems.' This is expected to enhance the system's performance (Jaque et al., 2014).

8.3.3. SAR and Lidar Systems

Currently, there is limited research on the use of SAR and Lidar remote sensing in monitoring pests and plant diseases. This could be because there is a weak correlation between the pest and plant disease symptoms and the SAR and Lidar variables or parameters. Using SAR and Lidar, it is possible to analyze various characteristics of plants alongside their surroundings. This information can be valuable for monitoring diseases and pests. Cloud cover hinders the use of VIS-NIR RS pictures because areas with rainy and cloudy weather are conducive to a lot of pests and plant diseases. With the SAR cloud penetrating potential, monitoring can be conducted in any weather condition (Singh et al., 2007). This is a valuable aspect of remote sensing that allows for active monitoring. Utilizing the advancements in high-resolution and full polarimetric SAR technology, researchers have successfully employed SAR in extracting valuable information such as soil characteristics, plant water contents, and structural parameters of plant canopies. This has proven particularly effective for tall plants like sugarcane, corn, and trees. These observations have the potential to be utilized in identifying the habitat features of pests/plant diseases. In a study conducted by Singh et al. (2007), they found a correlation between different polarizations (VV- and HH-) and X-band incident angles (9.5 GHz) with the occurrence of pests in a sorghum field. This suggests that microwave remote sensing data could be used for pest monitoring purposes. Lidar can provide precise data on the structure and shape of the canopy. Through the use of Lidar sensors, the data collected on plant canopies can be transformed into a detailed 3D structure. This valuable information allows for the identification of any loss of biomass or damages caused by pests or diseases. Plant diseases, like sclerotinia blight in peanuts, can greatly change the appearance of canopy morphology in plants (Butzler et al., 1998). Lidar sensors may have a chance to identify such diseases. It is possible to discern their impact and occurrence by examining Lidar temporal series data throughout a crop season. Research has shown that SAR and Lidar RS observations can be valuable additions to VIS-NIR observations when it comes to monitoring pests/plant diseases (Martinelli et al., 2015).

8.3.4. Various Scales to Monitor Pests/Plant Diseases

When examining studies on the use of RS to track pests/plant diseases, it is important to consider the different scales at which these studies were conducted. These scales typically include regional, canopy, plot, and leaf levels. Most studies at the canopy of the leaf scales have been conducted via supervised experiments. The goal of the research is to:

- Comprehend how pests or diseases respond to remote sensing.
- To determine reliable and sensitive remotely sensed features.

Leaf-level monitoring offers a clear symptom signal from pests or diseases, aiding in the comprehension of their spectral responses. Spectral signals influenced by canopy morphology and structure would complicate studies at the canopy level. When identifying the impact caused by the rice leaf folder insect, Huang et al. (2012) carefully chose specific spectral indices and bands at the canopy and the leaf levels. Although there is some overlap between the most responsive indices and bands at the canopy and leaf levels, it was found that there are fewer indices and bands recognized at the canopy level. There is a related pattern that can be observed in the detection of damage due to spider mites in peaches, as mentioned in a study by Luedeling et al. (2009). Sensing at the canopy level may offer valuable insights that could be easily correlated with information collected by the satellite or airborne remote sensors. When observing powdery mildew in winter wheat, it was found that the spectral curves of healthy samples; and severely and mildly infected samples displayed a uniform pattern at both the canopy and leaf level. There is a strong correlation between the spectral responses of canopy-level diseases and those at a plot level, as shown by the spectral ratio curves of the initial bands and various VIs.

Studies conducted at the regional and plot level have placed a greater emphasis on the application of remote sensing features and mapping applications. With the help of advanced remote sensing systems, such as MAV or UAV systems or Hi-Res satellite systems, accurate mapping of damage at the plot level may be achieved. This can greatly assist in the precise application of pesticides or fungicides in the field, improving crop management. Based on regional mapping, the accuracy of mapping at this level is typically low due to the landscape complexity. For example, Zhang et al. (2015) reported an accuracy of 78%. Competent monitoring across sizable regions is highly valuable for regional-scale plant protection. Given the variation in the background effects at various scales, it becomes crucial to assess the reliability of remote sensing attributes and recalibrate models during the upscaling process.

8.4. FEATURES OF RS FOR MONITORING PESTS/ PLANT DISEASES

To effectively monitor pests/plant diseases using remote sensing observations, it is essential to identify distinct and effective features. Until now, numerous remote sensing features have been suggested or recognized to detect plant symptoms resulting from pests and diseases, as well as to map their habitats. The primary features include image-based and landscape-based features like the fluorescence/thermal features; and the spectral VIS-NIR features (Table 8.2). Considering the diverse range of damage mechanisms caused by pests and diseases, it is often essential to conduct a sensitivity evaluation of remote sensing features to ensure effective monitoring (Cock et al., 2016) (Figure 8.2).

Table 8.2. *Summary of Crops and the Corresponding Plant Diseases Detected by Spectroscopy or Hyperspectral Techniques*

Crop Types	**Crop Names**	**Disease Names**	**Disease Types**	**Main Infected Sites**	**Sensors/ Platforms/ Scales**	**Analysis Approach**
Cereal crops	Wheat	Wheat stripe rust	Fungal diseases	Leaves *	Analytical spectral device (ASD)/ handheld/ leaves and canopy	Partial least squares discriminant analysis (PLSR), support vector regression (SVR), and Gaussian process regression (GPR).
		Wheat leaf rust	Fungal diseases	Leaves *	ASD and digital camera/handheld/leaves	Linear spectral mixture analysis, Fisher function.
		Wheat powdery mildew	Fungal diseases	Leaves *	ASD/handheld/canopy	Continuous wavelet analysis, Fisher's linear discrimination analysis (FLDA), and support vector machine (SVM).
		Fusarium head blight	Fungal diseases	Ears * and stalks	ImSpector V10E and ImSpector N25E/indoor measurement platform/ spikelets	Linear model fitting, spectral vegetation indices (SVIs).
	Rice	Rice sheath blight	Fungal diseases	Leaves *	ImSpector V10E/indoor measurement platform/single plant	Linear discriminant analysis (LDA) and SVM.

		Rice blast	Fungal diseases	Leaves *, stems, and ears	ORCA-05G/ darkroom/ panicle	"Bag of spectra words" (BoSW) model and chi-square support vector machine (chi-SVM).
	Maize	Grey leaf spot disease	Fungal diseases	Leaves *	ASD and three multi-spectral satellite Resampled/ handheld/ satellite/ leaves and canopy	Random forest algorithm (RF).
		Leaf spot disease	Fungal diseases	Leaves * and bracks	ASD/hand-held/leaves	Guided regularized random forest (GRRF) and RF.
		Ear rot	Fungal diseases	Ears and kernels *	SisuChema/ HgCdTe detector/fungal isolates	Principal component analysis (PCA) and PLSR.
Legume crops	Soybean	Soybean anthracnose	Fungal diseases	Stems *, pods, and leaves *	Pika XC/ mounting tower/stems	Genetic algorithm as an optimizer and SVM as a classifier.
		Yellow mosaic virus	Viral disease	Leaves *	ASD/hand-held/leaves	Spectral derivative and red edge analysis.
Tuber crops	Potato	Late blight disease	Fungal diseases	Leaves * and fruits	Rikola/ unmanned aerial vehicle (UAV)/ plots	Simplex volume maximization (SiVM) and pixel-wise log-likelihood ratio (LLR) calculation.
		Potato virus Y	Viral disease	Leaves *	Specim FX10/tractor/canopy	Deep learning, fully convolutional neural network.

Sugar crops	Sugar beet	Cercospora leaf spot	Fungal diseases	Leaves *	ASD/handheld/leaves	Spectral signature analysis and vegetation indices.
		Beet rust	Fungal diseases	Leaves *	ImSpector V10E/ microscope/ tissue	Spectral angle mapper (SAM).
		Beet powdery mildew	Fungal diseases	Leaves *	ASD/leaf clip/leaves	SVIs and SVM.
		Root rot	Fungal diseases	Roots (leaves)*	ASD/handheld/canopy	SVIs and nonlinear regressions.
Vegetables	Tomato	Gray mold	Bacterial diseases	Fruits, leaves * and stems	ImSpector V10E/indoor measurement platform/leaves.	K-nearest neighbor (KNN), C5.0 models, and feature rank.
		Tomato yellow leaf curl virus	Viral disease	Leaves *	Imspector V10E-QE/ indoor measurement platform/ leaves.	Grey level co-occurrence matrix (GLCM).
Fruits	Citrus	Citrus canker	Bacterial diseases	Fruits * and leaves *	Pika L 2.4/ mounting tower and UAV/leaves, fruits, and single plant	Radial basis function (RBF) and KNN.
		Huanglongbing (citrus greening)	Bacterial diseases	Fruits *, leaves * and roots	AISA Eagle/ airborne/ canopy	SVM

Source: Zhang, N., Yang, G., Pan, Y., Yang, X., Chen, L., & Zhao, C. (2020). A review of advanced technologies and development for hyperspectral-based plant disease detection in the past three decades. Remote Sensing, 12(19), 3188. https://doi.org/10.3390/rs12193188.

* Indicates the main organs that hyperspectral-based studies focus on.

* Indicates sugar beet root rot which mainly infects the roots, but the main organ that hyperspectral-based studies focus on is the leaves.

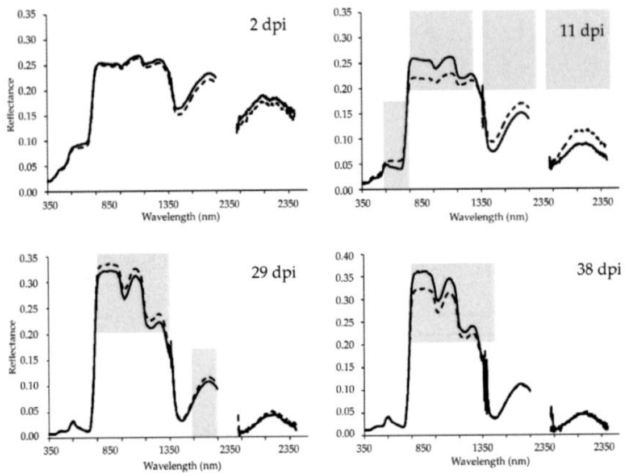

Figure 8.2. *Effect of the yellow rust development on the spectral reflectance of winter wheat as days post inoculation (dpi) increase. Gray areas in the reflectance images (lower) indicate significantly different spectral ranges between the inoculated (dotted black line) and non-inoculated (solid black line) leaves.*

Source: Zhang, N., Yang, G., Pan, Y., Yang, X., Chen, L., & Zhao, C. (2020). A review of advanced technologies and development for hyperspectral-based plant disease detection in the past three decades. Remote Sensing, 12(19), 3188. https://doi.org/10.3390/rs12193188 (Creative Commons License).

8.4.1. Spectral VIS-NIR Features

Given the widespread use of VIS-NIR sensors to monitor pests and plant diseases, it is common to rely on the spectral features provided by these systems. One of the most straightforward features is band reflectance. Several studies have found that certain spectral regions, such as NIR, red, and green, are highly responsive to different pests and plant diseases. There are various methods to transform spectral reflectance values, including continuous wavelet (CWT); continuous removal (CRT); and spectral derivatives transformations (SDT). To improve the ability to capture infection symptoms, these spectral transformations have also been used, albeit in smaller quantities. Typically, SDT is used to extract the convex properties (such as red-edge amplitude and position) from plant spectra. In the study, it was observed that rice plants infested by rice leaf folder showed a significant blue shift of the red-edge peak, according to their 1st derivative value.

Using the CRT, a method has been developed to accurately identify concave-convex shape characteristics across specific absorption positions throughout a spectrum (Ramirez et al., 2013). This technique has shown promising results in detecting wheat powdery mildew and the tomato lead miner. Continuous wavelet analysis or CWA has recently been developed as a new technique for identifying pests/plant diseases. Due to its ability to record very subtle changes in the spectral shape, the CWA-derived features have been found to outperform conventional spectral features in the winter wheat to detect yellow rust (Rossel et al., 2009) (Figure 8.3).

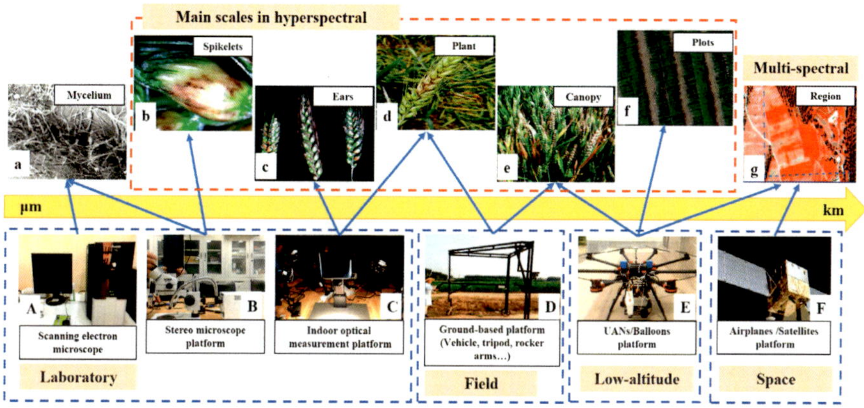

Figure 8.3. *Platforms and scales commonly used in plant disease research employing hyperspectral techniques, with wheat Fusarium Head Blight (FHB) as the example. (a–c) represent indoor measurement platforms at different scales: a scanning electron microscope, a stereo microscope platform, and a typical indoor measurement platform, respectively. (d) shows a field vehicle equipped with a hyperspectral imaging system. (e) depicts an unmanned aerial vehicle (UAV) platform. (f) illustrates a satellite platform. Images (a–g) display hyperspectral images of FHB-infected winter wheat at various scales, indicating the presence of disease mycelium, infected spikelets, ears, plants, canopy, plots, and region hyperspectral images, respectively.*

Source: Zhang, N., Yang, G., Pan, Y., Yang, X., Chen, L., & Zhao, C. (2020). A review of advanced technologies and development for hyperspectral-based plant disease detection in the past three decades. Remote Sensing, 12(19), 3188. https://doi.org/10.3390/rs12193188.

In addition to analyzing the amplitude of the transformed spectral bands and reflectance, there are various VIs that are commonly employed to monitor and detect pests and plant diseases. By utilizing algebraic combinations of original bands, VIs have the ability to highlight significant functionality or biophysical properties of the plants (Table 8.3). Given the relationship between the spectral responses and plant pigment content, several vegetation indices are utilized to detect pests and plant diseases that impact pigments like chlorophyll, carotenoid, anthocyanin, carotenoid, and chlorophyll; and pigment ratios such as the xanthophyll cycle, and Car/Chl. Furthermore, certain vegetation indices

(VIs) that are influenced by various plant parameters or overall plant health have been discovered to be highly responsive to a range of diseases and pests (Dotto et al., 2017). These VIs include the Optimized Soil-Adjusted Vegetation Index or OSAVI; Triangular VI or TVI; Green NDVI or GNDVI; and Normalized Difference Vegetation Index or NDVI. Several studies have demonstrated the sensitivity of these VIs to different diseases and pests. Furthermore, in order to identify the infestation of leaf miners in tomatoes, Xu et al. (2007) utilized a WBI as a means of detecting plant wilting caused by pests or diseases. In a study conducted by Naidu et al. (2009), it was discovered that the water index or WI showed a high level of sensitivity to leafroll disease in grapevines. In addition, new vegetation indices were specifically developed to detect certain pests or diseases. Apan et al. (2004), for instance, created the disease-water stress index or DSWI to identify sugarcane orange rust. In a study conducted by Luo et al. (2013), they introduced an innovative aphid index (AI) for identifying aphid infestations. Prabhakar et al. (2011) introduced a leaf hopper index or LHI that relies on specific bands for identifying leafhoppers in cotton.

8.4.2. Fluorescence and Thermal Parameters

When it comes to detecting pests and diseases using thermal or fluorescence sensors, certain parameters have been developed or introduced to establish a connection between the infection symptoms and signals. Several studies have utilized the ratio of fluorescence amplitudes at specific peaks in the continuous fluorescence spectra to pre-symptomatically detect pathogens. Also, using saturation pulse fluorescence analysis, many fluorescence parameters were used to find plant diseases and pests. These included the effective quantum yield of photosystem II or ΦPSII; non-photochemical □uenching or NPQ; the max efficiency of PSII photochemistry in light-adapted material given by $F_{v'}/F_{m'};$ primary photochemistry given by F_v/F_m; and the maximum quantum efficiency of photosystem II (PSII). When it comes to identifying pests and plant diseases, features obtained from the thermal infrared sensors are more straightforward and much simpler compared to fluorescence or spectral features (Permyakov & Burstein, 1984). In essence, the most significant signal is the temperature of the canopy or leaves. Based on research conducted in a greenhouse environment, it was discovered that the temperature difference between the leaf and the air can be used as a diagnostic tool for identifying pre-symptomatic downy mildew disease infection in grapevines. In a study conducted by Falkenberg et al. (2007), it was shown that an infrared camera can be used to create a canopy temperature map that effectively detects early signs of cotton root rot infection (Chaerle et al., 2007). Considering that fluorescence, thermal, and optical observations can offer valuable insights into infection symptoms, combining these approaches is anticipated to improve stress detection. For example, in the study conducted various thermal; fluorescence; and optical features were utilized to detect verticillium wilt in olives. The objective was to accomplish prompt identification of the disease and accurately differentiate between different infection levels (Bushueva et al., 1978).

8.4.3. Landscape and Image-Based Features

With the use of various remote sensing systems, such as thermal, fluorescence, and VIS-NIR, valuable observations can be obtained and analyzed to effectively monitor pests and plant diseases. The image-based analysis allows for the precise mapping of infection areas and the estimation of severity based on the landscape or texture features (Figure 8.4). Texture features, such as uniformity, mean intensity, correlation, inverse difference, entropy, and contrast among other variables, play a crucial role in detecting pests and plant diseases, especially at a micro (leaf) level. The color co-occurrence method (CCM) is a widely used technique for extracting these features. Pydipati et al. (2006) utilized 13 CCM-based features and successfully identified grape leaves that were infected by greasy spots, melanose, and scabs (Laliberte et al., 2007). Their classification accuracy exceeded 95%. Yao et al. (2009) achieved an impressive accuracy of 97.2% in classifying rice blast, rice-sheath blight, and rice bacterial leaf blight using a set of texture and shape features. Furthermore, the inclusion of image texture features, along with spatial metrics derived from remote sensing images, can effectively detect, and analyze the spatial distribution patterns of pests and plant diseases. These insights are particularly valuable for managing and monitoring crop health at a larger scale, such as at the plot or regional level.

Considering the unique spatial patterns of pest and disease infestations, as opposed to other plant stressors like drought, various landscape metrics can be employed to differentiate them. These metrics include class area, shape index, contiguity index, proximity index, clumpiness index; proximity, contiguity, and shape index; and class area (Iqbal et al., 2020). One interesting example involves distinguishing between Russian wheat aphids and other stresses such as agronomic and drought conditions. When it comes to quantifying the spatiotemporal distribution of the bark beetle infestations, it is important to consider the various perspectives that image-based features offer in the monitoring of pests and plant diseases. In addition to thermal, fluorescence, and optical features, these image-based features provide valuable insights. Utilizing various thermal, fluorescence, and optical features, valuable information can be extracted to analyze landscape or textural characteristics. This approach effectively combines different features to monitor and identify pests and plant diseases (Li, 2023).

8.4.4. Features Related to Habitat Characteristics

Furthermore, there have been efforts to utilize remote sensing data to identify the ideal habitats for pests and plant diseases, in addition to its direct application in assessing damages caused by them. Certain parameters derived from the Tasseled Cap Transformation (TCT) can serve as effective indicators of habitat suitability. These parameters, such as wetness, brightness, and greenness, are closely linked to soil moisture and plant vigor. It has been observed that stressed plants are more susceptible to pests and diseases, making these TCT-based parameters valuable in assessing habitat conditions (Willby et al., 2000). In addition to the TCT-based parameters, the use of LST and VIs is crucial for assessing habitat conditions. Several vegetation indices, such as NDWI, Wi, MSAVI 2, MSI, PRI, SAVI, and TVI are useful in monitoring the presence of wheat powdery mildew; tree wooly adelgid; spruce budworm; and larval mosquito. There are satellite imagery products, such as MODIS-LST, that provide information on transpiration intensity in plants. Additionally, thermal bands from satellites like HJ-IRS, ASTER, and Landsat TM can be used to retrieve this data. Yuan et al. (2017) extracted LST, TCT, and VI parameters from LandSat 8 and WorldView 2 satellite images to assess the environmental conditions and the wheat growth status, specifically focusing on aphid and powdery mildew monitoring. With a validation accuracy of 82%, it is possible to effectively monitor the suitable habitat for pests and diseases. In addition, considering the complexity of habitat characteristics, supplementary data such as solar radiation, terrain attributes, and certain meteorological parameters (e.g., rainfall, humidity, and temperature) can be combined with remote sensing features to monitor habitats. In a study, a disease forecasting system was developed by integrating remote sensing features with meteorological parameters. The results showed that the inclusion of remote sensing data significantly improved accuracy, increasing it from 69% to 78% (Southwood, 1977).

8.4.5. Sensitivity Analysis for the Features Selection

Considering the diverse habitat characteristics and symptoms caused by pests and plant diseases, it is crucial to perform a sensitivity analysis to determine the most suitable remote sensing features to monitor them. To accomplish this, certain statistical analyses were utilized to evaluate the sensitivity of various features. For instance, when classifying pests and disease types, or differentiating infection levels, researchers often relied on the Analysis of Variance (ANOVA) and independent t-tests (Jeanneret et al., 2003). The Pearson correlation analysis is a commonly used method for conducting sensitivity analysis to identify remote sensing features. Furthermore, the evaluation of feature sensitivity can be done using the predictive residual sum of squares or PRESS information of the PLS regression. This is particularly useful when there is a high level of collinearity among features. In addition, a feature optimization strategy called the RELIEF-F algorithm is also presented to identify features that demonstrate strong robustness. For example, Mahlein et al. (2016) utilized the RELIEF-F algorithm to identify a set of reliable bands that effectively distinguish powdery mildew, leaf rust, and leaf spot in sugar beet. This approach resulted in an impressive classification accuracy of over 85% for all classes (Hassan et al., 2023).

Given the dynamic nature of symptoms caused by pests and diseases, it is crucial to identify appropriate features at various stages. When it comes to detecting wheat yellow rust, a comprehensive analysis was conducted by Zhang et al. (2015) to evaluate the sensitivity of various spectral features throughout five distinct phases. At various phases, distinct feature sensitivity patterns were noticed. Researchers have found that there is a significant shift in the sensitive band when detecting rice leaf folder. At the tillering stage, the band is at 757 nm, but it shifts to 445 nm at the heading stage.

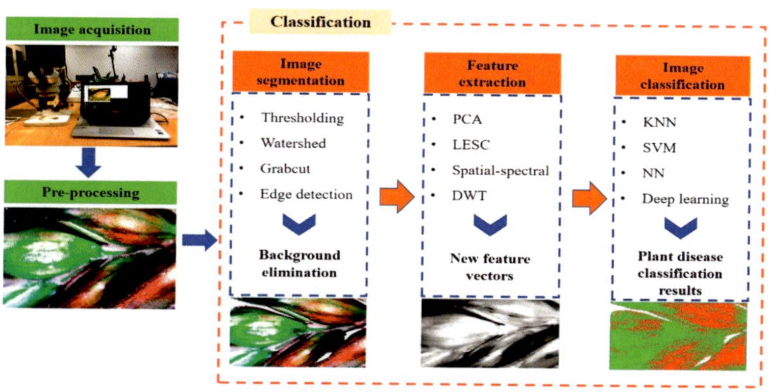

Figure 8.4. *Basic workflow in hyperspectral imaging-□ased plant disease classification. [PCA: principal component analysis; LESC: local em□edding □ased on spatial coherence algorithm; DWT: discrete wavelet transforms; KNN: K-nearest neigh□or; SVM: support vector machine; and NN: neural networks].*

Source: Zhang, N., Yang, G., Pan, Y., Yang, X., Chen, L., & Zhao, C. (2020). A review of advanced technologies and development for hyperspectral-based plant disease detection in the past three decades. Remote Sensing, 12(19), 3188. https://doi.org/10.3390/rs12193188.

In addition, it is important to acknowledge that the feature sensitivity may differ depending on the scale of observation. A study conducte d by Huang et al. (2012) on the monitoring of rice leaf folder found that there are seven sensitive spectral regions at a leaf level (722–770, 703–715, 688–699, 581–606,

550–568, 526–545, and 503–521 nm), while at a canopy-level, only one region (747–754 nm) was found to be sensitive. According to a study by Luedeling et al. (2009), two spectral regions were identified at the canopy level and eight regions at the leaf level using PLS-PRESS.

8.5. TECHNIQUES FOR MONITORING PLANT DISEASES AND PESTS

8.5.1. Algorithms for Monitoring Plant Pests and Diseases

To effectively utilize the RS features derived from various types of RS data to monitor pests and plant diseases, a range of algorithms (or combinations) can be employed or created to facilitate the following tasks.

- Identifying a particular pest or disease;
- Specializing in distinguishing pests and diseases;
- Determining the degree of infection.

This section will review current algorithms for pest and disease monitoring, emphasizing the state of research in each area (Table 8.4).

8.5.1.1. Statistical Discriminant Analysis

In certain cases where we need to detect a specific disease or differentiate between pests and diseases, statistical discriminant analysis methods have shown remarkable accuracy. In a study conducted stepwise discriminant analysis had been employed to identify grapevines that were affected by *Vitis vinifera* L. This analysis utilized a combination of selected bands and vegetation indices, resulting in an impressive overall accuracy of 81%. Yuan et al. (2014b) developed a discriminant model using Fisher's linear discriminant analysis or FLDA to distinguish between aphids, powdery mildew, and yellow rust in winter wheat. The discriminant model achieved a total accuracy of 75%. Furthermore, studies have shown that logistic regression may effectively identify apple scab (Delalieux et al., 2007) and sugar beet rhizomania (Steddom et al., 2003).

8.5.1.2. Regression-Based Models

Regression-based methods are commonly employed to assess the severity of pests and plant diseases. In a study conducted by Huang et al. (2007), linear regression was used to measure the yellow rust disease index or DI in wheat. According to Singh et al. (2007), a high-order polynomial regression model was able to accurately identify the levels of infestation damage by sorghum grasshoppers using a microwave-derived parameter. Instead of using univariate regression models, experts in the field often rely on multivariate regression models like PLSR or MLR to monitor pests and plant diseases. These models utilize different features and have been extensively studied by researchers. When comparing PLSR to traditional MLR, one can see that PLSR offers an advantage. It not only reduces the raw data dimensionality but also preserves a significant portion of its variance. It was discovered that PLSR outperformed MLR in accurately determining the severity of bacterial spots in tomatoes caused by *Xanthomonas perforans*. In a study conducted a similar pattern was discovered when estimating the extent of wheat powdery mildew (Regmi et al., 2014).

8.5.1.3. ML and Deep Learning Algorithms

Several ML algorithms are crucial in modeling to identify pests and plant diseases. For instance, the detection of head blight disease in wheat can be achieved using spectral angle mapping (SAM) and hyperspectral imaging systems in combination. Additionally, the monitoring of wheat powdery mildew has been effectively carried out using a high-resolution SPOT-6 image. With the utilization of a Quickbird image and the application of MLC successfully mapped the damage caused by the mountain pine beetle with an impressive accuracy rate of 91% (Balyen & Peto, 2019).

Table 8.3. *List of Major Contributions According to Different Methods of Hyperspectral Image Classification for Plant Disease Detection, Identification, and Mapping*

Plant and Diseases	Targets *	Scales	Methods and Algorithms	Classification Accuracy
Sugar beet and *Cercospora leaf spot*/ powdery mildew/ sugar beet rust	Disease identification	Leaf	Spectral angle mapper (SAM)	98.9% for *Cercospora* leaf spot at 8 dai; 97.23% for powdery mildew at 14 dai; 61.70% for sugar beet rust at 20 dai.
Wheat and *Fusarium* head blight	Disease identification	Spike	Support vector machine (SVM) with reflectance and spectral vegetation indices (SVIs)	95.0% and 99.0% for two classes of classification using SVIs and reflectance; 76.0% and 77.0% for multiclass classification using SVIs and reflectance.
Wheat and yellow rust (*Puccinia striiformis*)	Disease detection	Leaf	Quadratic discriminant analysis (QDA)/ self-organizing map (SOM) NN	94.5% by using QDA; and around 99% by using SOM NN.

Citrus and citrus bacterial canker	Disease severity classification (asymptomatic, early, and late symptoms)	Leaf/fruit/plant	Neural network radial basis function (RBF); KNN with SVIs.	94%, 96%, and 100% by RBF and 94%, 95%, and 96% by KNN for three levels at leaf scale; 92% canker detection at fruit scale and 100% and plant scale.
Soybean and charcoal rot	Disease identification	Stem	Three-dimensional convolutional neural network (3D CNN)	95.73%
Wheat and stripe rust	Disease identification and mapping	Canopy/plot	Linear regression model	–

Source: Zhang, N., Yang, G., Pan, Y., Yang, X., Chen, L., & Zhao, C. (2020). A review of advanced technologies and development for hyperspectral-based plant disease detection in the past three decades. Remote Sensing, 12(19), 3188. https://doi.org/10.3390/rs12193188.

* The targets of classification are divided into disease detection, disease identification, disease severity classification, and disease mapping; the authors used SVIs fitting the linear regression model and then mapped disease severity at the plot scale by model inversion, so there was one classification accuracy.

Furthermore, certain unsupervised classification techni□ues such as Iterative Self-Organizing Data Analysis (ISODATA) and nonparametric techniques like random forest (RF) are also used in the surveillance of pests and plant diseases (Janiesch et al., 2021).

Support vector machines or SVMs are also extensively used in the monitoring of various pests and plant diseases, showcasing their effectiveness among ML algorithms. Using SVM classifiers with fluorescence features, Römer et al. (2011) succeeded in the early identification of wheat leaf rust with an impressive accuracy of 93%.

When it comes to detecting Huanglongbing (HLB) in citrus, SVMs have proven to be more effective than other methods such as neural networks, logistic regression, KNN, decision trees, QDA, and LDA. Studies have demonstrated that SVMs outperform SAM and Random Forest when it comes to accurately mapping wheat damage caused by powdery mildew and pest defoliation. With its ability to find a separating surface through optimization, SVM proves to be highly effective in handling limited training samples. This is particularly useful in remote sensing monitoring works for pests and plant diseases, where the availability of training data is often a challenge (Abdel et al., 2022).

In certain complex situations, Deep learning approaches along with their ancestor, artificial neural networks, or ANN, have shown a remarkable ability to monitor pests and plant diseases. Through the use of optimized geometric and spectral parameters, Castro et al. (2015) successfully detected laurel wilt disease in avocados at an early stage using a feed-forward back propagation (BP) neural network. In addition to traditional neural network models, recent advancements in deep learning models have been utilized in the agricultural field, as highlighted by Singh et al. (2018). These state-of-the-art models hold significant promise for effectively recognizing and diagnosing plant stresses. Ferentinos (2018) used Convolutional Neural Networks or CNNs to detect and diagnose plant diseases with an accuracy of 99.53% based on a huge number (n = 87,848) of leaf photos. It is important to note that the need for a significant amount of labeled samples is a major obstacle in the practical use of many deep learning methods. Collecting a substantial number of training samples for pests and plant diseases from remote sensing data is crucial for future studies on the practicality of deep learning methods (Naskath et al., 2023).

8.5.1.4. Spectral Unmixing Algorithms

In the field of pest and plant disease monitoring, spectral unmixing methods are gaining popularity. These methods involve assuming that the infected plant represents a specific end member. In a study conducted a constrained linear spectral mixture analysis was used to map damage in cotton fields caused by spider mite. A big problem is figuring out what the reference endmembers are for infected plants, healthy plants, or plants in some other state (Pu et al., 2013). For space-borne images, the reference endmembers spectra typically originate directly from the images so that the in-situ spectra used by the spectrometer do not compare incorrectly with the image-derived spectra. In the study conducted advanced techniques such as the pixel purity index (PPI) and minimum Noise Fraction (MNF) transform were utilized to accurately identify the endmember spectra of tomato plants affected by late blight (Keshava & Mustard, 2002).

The mixture-tuned matched filtering (MTMF) method is particularly intriguing due to its advanced spectral unmixing algorithm. One of its notable advantages is that it can be used even when the materials (end members) in the image scene are not all known. Monitoring pests and plant disease stresses can be advantageous, as their impacts often blend with other factors such as nutrient stresses and vigor differences, making them difficult to identify. Using a Quickbird image successfully utilized the MTMF technique to accurately map disease severity in a wheat field. In a study conducted by Zhang et al. (2015), it was demonstrated that MTMF proved to be successful in accurately mapping regional disease using HJ-CCD satellite moderate-resolution images. The authors highlighted the potential of combining the MTMF PLSR and MTMF techniques to enhance the accuracy of wheat powdery mildew monitoring (Heylen et al., 2011). By implementing this combination, they were able to achieve a significant improvement in accuracy, increasing it from 72% to 78%.

8.5.2. Employing Single-Phase or Multi-Phase RS Images for Mapping

When it comes to monitoring pests and plant diseases, mapping with single-phase and multi-phase remote sensing images are two crucial approaches (Buetehorn et al., 2011).

8.5.2.1. Mapping with Single-Phase RS Image

One important premise to consider is that the damage caused by disease/pests often exhibits distinct and highly contrasting features at a specific stage. Studies in this field typically rely on data collected from aerial or satellite imagery. According to a study by Yang et al. (2010), when a disease symptom has a distinct spectral response, like cotton root rot, multispectral, and hyperspectral remote sensing images were found to be effective for monitoring. In addition, it is worth noting that certain new channels, like the red-edge channel found in advanced satellite sensors such as Rapideye, and WorldView 2, 3 have the potential to enhance monitoring capabilities.

It was discovered that the red-edge band and its corresponding indices, developed using the single-phase WorldView-2 image, proved to be highly effective in accurately mapping the extent of damage due to the forest bronze bug. Considering the rapid spread of pest infestations or disease infections to the full canopy, it is crucial to carefully choose the optimum time for monitoring. As mentioned by Yuan et al. (2017), the mapping of wheat powdery mildew was found to be most effective during the grain-filling stage. In addition, having beforehand knowledge of the mapping area is crucial for excluding other potential stresses such as pests, diseases, and nutrient stresses when conducting a single-phase mapping (Kushwaha et al., 2000).

8.5.2.2. Mapping with RS Images at Multiple Phases

In certain situations where the effects of pests or diseases can be easily mistaken for other factors, a single-phase remote sensing image may not provide enough information for accurate mapping. On the other hand, it is crucial to utilize multi-phase RS images that capture temporal features and provide information about the spread of pest attacks or disease infections for effective monitoring (Figure 8.5). There are three common methods to perform multi-phase mapping:

i. Mapping using images from two phases;
ii. Mapping using images from multiple phases, often three to five; and
iii. Mapping upon a particular temporal trajectory (Witoszynskyj et al., 2009).

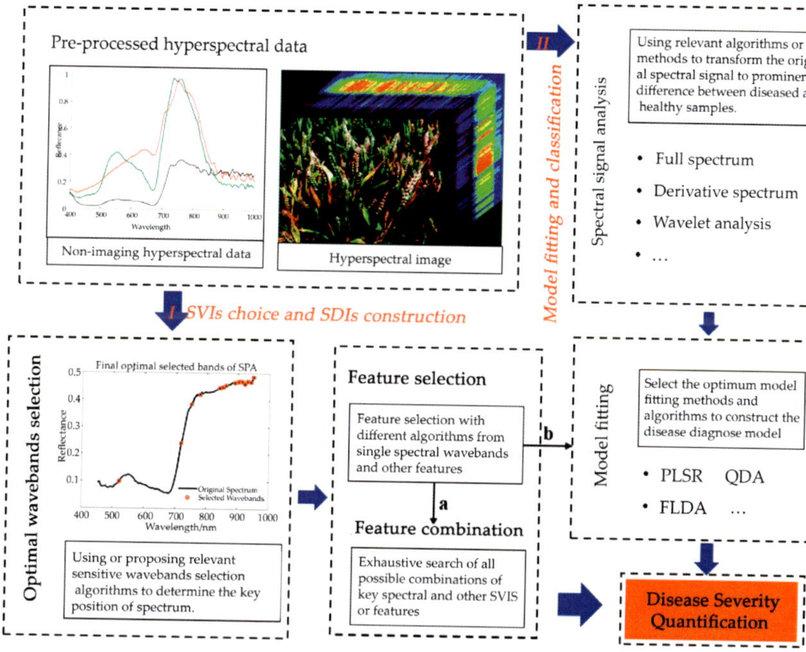

Figure 8.5. *Two □asic workflows of disease severity quantification. (a) and (□) indicate that after optimal waveband and feature selection, there are two methods of constructing the final SDIs, one □ased on feature com□ination and the other involving model fitting □y statistical analysis methods. [SVIs: spectral vegetation indices; SDIs: special disease indices; PLSR: partial least squares regression; QDA: quadratic discriminant analysis; and FLDA: fisher's linear discrimination analysis].*

Source: Zhang, N., Yang, G., Pan, Y., Yang, X., Chen, L., & Zhao, C. (2020). A review of advanced technologies and development for hyperspectral-based plant disease detection in the past three decades. Remote Sensing, 12(19), 3188. https://doi.org/10.3390/rs12193188 (Creative Commons License).

Typically, the two-phase monitoring is implemented when there is an outbreak of pest infestation or plant disease. For this scenario, it is recommended to capture one image before the outbreak event and another image during or after the peak damage. Several studies have highlighted the importance of this approach. Change detection analysis was carried out using two HJ-CCD images that were captured before and after the armyworm outbreak. The results indicated that the overall accuracy of damage mapping significantly increased from 50% to 79% when using two-phase mapping instead of single-phase mapping (Cusack & Papadakis, 2002).

In addition to using the two-phase mapping procedure, utilizing multi-date remote sensing images (such as 3–5 phases) can provide a wealth of valuable information for monitoring purposes. When analyzing the damage caused by the mountain pine beetle, conducted a study on changes in the Red-Green Index (ΔRGI) using Quickbird images. They found that the true positive accuracies ranged from 89 to 93% across four different times. In a study conducted by Zhang et al. (2015), high-resolution HJ-CCD images were utilized on four different dates to effectively monitor the occurrence and severity of wheat powdery mildew at the regional level. Through a thorough feature selection process, it was discovered that the utilization of multi-stage spectral features played a crucial role in reducing mapping uncertainties. As a result, a mapping accuracy of 78% was achieved.

Given the increasing availability of remote sensing images, it is now possible to extract temporal trajectories from time-series data. These trajectories can help identify the specific temporal patterns of infestations or infections. Remote sensing data with a higher revisit frequency, like SPOT-VGT and MODIS, have commonly been utilized for monitoring of this kind. When monitoring diseases or pests that have slower development progress, such as certain forest diseases or pests, it may be worth considering using remote sensing data with longer time intervals. As an experienced professional in the field of crop management and remote sensing, the work of Goodwin et al. (2008) is presented. They conducted a study where they utilized a series of Landsat TM data, consisting of 8 images taken over 14 years, to monitor the spread of mountain pine beetle infestation. This research provides valuable insights into the dynamics of this destructive pest. Several well-known vegetation indices, such as NDVI and EVI, are commonly used as proxies to analyze the temporal trajectory of vegetation. To guarantee the accuracy and dependability of the temporal trajectories, certain data preprocessing steps are required. These steps include curve smoothing; radiometric normalization; or spatial co-registration. For instance, the use of TIMESAT to process MODIS time-series data has been recommended by experts. Temporal features like change angles and magnitudes derived from temporal trajectory curves are utilized to characterize the temporal features of infestations or infections in monitoring studies (Homer et al., 1997).

8.5.3. Typical Framework to Monitor Pests or Plant Diseases

Due to the significant variations in host plants, development processes, and symptoms for pests and plant diseases, it is not possible to establish a universal remote sensing feature or monitoring model. Nevertheless, methodologies and strategies to monitor pests or plant diseases share certain common characteristics. After thoroughly examining the RS systems, their features, and algorithms, a comprehensive framework may be developed like the following. This framework can be used as a valuable reference for monitoring unknown and known pests or diseases. For a visual representation, please refer to the flowchart in Figure 8.6. Firstly, individuals must have a solid understanding of the specific pest or disease they are dealing with (Luvisi et al., 2016). This includes being aware of the type of damage it causes, its development process, and the environments in which it thrives. Secondly, when considering a specific application scenario, researchers must determine the appropriate monitoring scale and specify the necessary remote sensing data. This includes identifying the type of remote sensing systems (as shown in Table 8.1) and the required temporal and spatial resolution of the data. Thirdly, it is crucial to conduct a series of experiments to gather training data that will aid in the feature selection and the process of modeling.

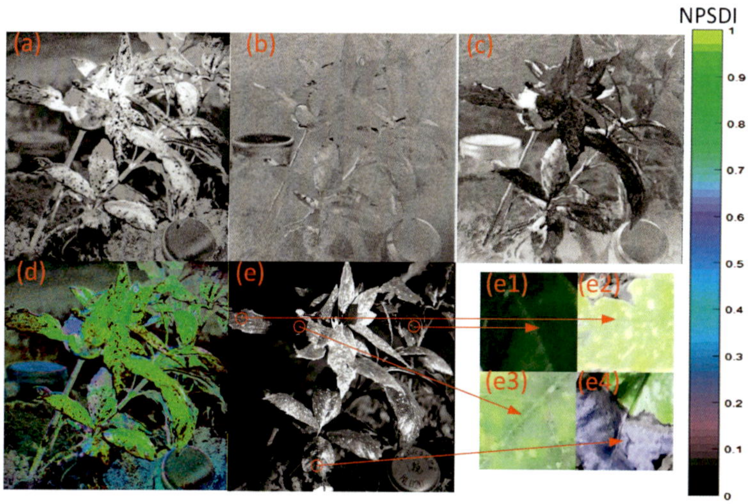

Figure 8.6. *(a) NDVI image; (□) AOP image; (c) DoLP image; (d) fusion image (NDAI); (e) gray-scale image; (e1) healthy leaf region; (e2) level-2 stress leaf region; (e3) level-1 stress leaf region; and (e4) withered leaf region.*

Source: Li, S., Jiao, J., & Wang, C. (2021). Research on polarized multi-spectral system and fusion algorithm for remote sensing of vegetation status at night. Remote Sensing, 13(17), 3510. https://doi.org/10.3390/rs13173510.

This approach emphasizes the importance of a hierarchical process in data collection (Andrade et al., 2003). It includes conducting experiments at the canopy or leaf level to gain a better understanding of how remote sensing responds to pests and diseases. Additionally, it is recommended to conduct regional scouting with synchronized image acquisition to ensure accurate results on a larger scale. If possible, it would be ideal to conduct the latter across multiple dates to take advantage of multi-phase modeling approaches. Additionally, it is important to determine the monitoring model algorithm and approach. Ultimately, it is crucial to conduct a performance assessment to evaluate the adequacy of the accuracy (Stenberg, 2017).

8.6. TRENDS AND CHALLENGES IN MONITORING PESTS AND PLANT DISEASE

Although significant progress has been made in monitoring pests and plant diseases in recent decades, there are still challenges that hinder the practical implementation of these techniques. Research focused on addressing these challenges will influence future trends. Table 8.4 presents an overview of UAVs in the monitoring of pests and plant diseases. It offers a detailed analysis of the challenges and trends associated with these issues (Kumar, 2014).

One of the primary concerns is identifying pests or plant diseases in their early stages. Considering the importance of timely monitoring for pests and plant diseases, it is crucial to have a reliable remote sensing system in place. However, it is often challenging to detect symptoms only when they are fully exhibited, as this may be too late to guide effective prevention measures. For better detection of pests or diseases in their early stages, it is crucial to explore the potential of fluorescence, Lidar, thermal, and SAR remote sensing observations (Balage et al., 2015). These can be combined with the already advanced VIS-NIR remote sensing observations. In addition, it is worth considering the use of multi-angular remote sensing to enhance the ability to detect lower canopy levels. The next challenge is accurately identifying a specific pest or disease in real-world field conditions, where multiple crop stresses may be present at the same time. Typically, monitoring applications or studies are carried out in controlled environments or areas with existing information, such as the specific pests and diseases or other stresses that have occurred in the field. When dealing with an area that is missing corresponding information, achieving accurate and reliable monitoring results can be quite challenging. It will be crucial to investigate the models' transferability and distinctive properties in more detail in the future. Cutting-edge algorithms, like deep learning algorithms, can be crucial in this process (Silva et al., 2021).

In addition, it is crucial to encourage the development of a comprehensive understanding of pests or diseases, including their climate conditions, types of soils, preferred habitats, and geographical distribution. This knowledge should be complemented by a network of pertinent ancillary data, such as soil and meteorological data, and information from wireless sensor networks. Having previous data can be extremely helpful in ruling out various possibilities and reducing uncertainty when monitoring pests and plant diseases in complex situations. The third aspect is the need to consistently monitor the progression of diseases or pests or diseases at a detailed level.

Table 8.4. *Advantages and Disadvantages of Using Drones for Agricultural Applications*

Drones	Advantages	Disadvantages
Rotary-wing (Multirotor)	• Take off and land vertically. • Return home capability. • High-detailed plant measurements. • Automatic recovery capability. • The lift is generated by batteries. • Hovering and flying at low altitudes. Depending on multiple rotors, for example, three rotors (tri-copters), four rotors (quadcopters), six rotors (hexacopters), and eight rotors (octocopters)	• Small payload sensor capacity. • Most of the capacity is taken by batteries. • Low endurance. • Low speed. • The area covered is limited.
Fixed wing	• The lift created by wings. • Longer flight. • High speed. • High endurance. • High payload sensor capacity. • Some have automatic recovery capability. • Fly a longer time and are more suitable for covering larger areas. • High altitude.	• Fly at an airspeed above their stall speed; sometimes problems occur in generating the desired data about the crops. • Low flexibility makes small crop monitoring very difficult. • Require runways and space to land and take off.

Hybrid vertical take-off and landing (VTOL)	• Resolve issues with multirotor and fixed-wing drones. • Combine fixed-wing drone cruise flight with multirotor drone VTOL capabilities. • Longer flight times, extensive coverage, quick vertical takeoffs, and minimal energy requirements.	• Expensive and hovering issues.

Source: Abbas, A., Zhang, Z., Zheng, H., Alami, M. M., Alrefaei, A. F., Abbas, Q., Naqvi, S. A. H., Rao, M. J., Mosa, W. F. A., Abbas, Q., et al. (2023). Drones in plant disease assessment, efficient monitoring, and detection: A way forward to smart agriculture. Agronomy, 13(6), 1524. https://doi.org/10.3390/agronomy13061524.

To accomplish this, the RS systems must possess a high resolution in terms of temporal, spectral, and spatial dimensions. At present, no remote sensing system can meet these requirements (Salehin et al., 2021). In addition, adverse weather conditions pose a significant challenge to the consistent collection of optical images. In order to address this problem, it is crucial to consider the potential of combining UAV images with Hi-Res satellite images to create continuous time-series remote sensing data. In addition, it is worth noting the significance of combining radar data with optical remote sensing data. The fourth aspect to consider is the sharing of information and data (Lamichhane et al., 2016). Given the widespread nature of pests and plant diseases, it is crucial to have international cooperation for research and practical implementation. Currently, the limited availability of survey data poses a challenge in the modeling of pests and plant diseases for monitoring purposes. It is recommended to gather data during the process of cultivation. For instance, it would be beneficial to involve extension officers and farmers in documenting the presence and intensity of pests and diseases in their cultivated fields using a smartphone application. It is crucial that the collected data is readily available to facilitate efficient data analysis and model training using advanced algorithms. It is anticipated that related observation networks and international projects will be established to facilitate experiments, data gathering, modeling, and the exchange of ideas on a continental or global level (Gullino et al., 2022).

8.7. SUMMARY

Monitoring pests and diseases over large areas is crucial for effective plant protection management and assessment in an efficient, timely, and reliable manner. Over the past few decades, a range of remote sensing techniques have emerged for monitoring pests and plant diseases. These techniques have shown great potential in complementing traditional inspection methods, which can be time-consuming and labor-intensive. The most recent advancements in applicable RS features and systems, monitoring algorithms, and feature selection methods, along with the models for carrying out effective and comprehensive monitoring of pests and plant diseases were compiled in this chapter. This overview of cutting-edge research findings in remotely sensed pest and disease monitoring is intended to spark new ideas and encourage the creation of related theories, procedures, and methods in both academic and industrial settings.

REFERENCES

1. Abdel-Jaber, H., Devassy, D., Al Salam, A., Hidaytallah, L., & El-Amir, M. (2022). A review of deep learning algorithms and their applications in healthcare. *Algorithms, 15*(2), 71.
2. Andrade, A. A., Lima, S. M., Pilla, V., Sampaio, J. A., Catunda, T., & Baesso, M. L. (2003). Fluorescence □uantum efficiency measurements using the thermal lens technique. *Review of Scientific Instruments, 74*(1), 857–859.
3. Apan, A., Held, A., Phinn, S., & Markley, J. (2004). Detecting sugarcane 'orange rust' disease using EO-1 Hyperion hyperspectral imagery. *International Journal of Remote Sensing, 25*(2), 489–498.
4. Balage, J. M., e Silva, S. D. L., Gomide, C. A., de Nadai Bonin, M., & Figueira, A. C. (2015). Predicting pork quality using Vis/NIR spectroscopy. *Meat Science, 108,* 37–43.
5. Balyen, L., & Peto, T. (2019). Promising artificial intelligence-machine learning-deep learning algorithms in ophthalmology. *The Asia-Pacific Journal of Ophthalmology, 8*(3), 264–272.
6. Barilero, T., Le Saux, T., Gosse, C., & Jullien, L. (2009). Fluorescent thermometers for dual-emission-wavelength measurements: Molecular engineering and application to thermal imaging in a microsystem. *Analytical Chemistry, 81*(19), 7988–8000.
7. Barzman, M., Bàrberi, P., Birch, A. N. E., Boonekamp, P., Dachbrodt-Saaydeh, S., Graf, B., & Sattin, M. (2015). Eight principles of integrated pest management. *Agronomy for Sustainable Development, 35,* 1199–1215.
8. Buetehorn, S., Volmering, D., Vossenkaul, K., Wintgens, T., Wessling, M., & Melin, T. (2011). CFD simulation of single-and multi-phase flows through submerged membrane units with irregular fiber arrangement. *Journal of Membrane Science, 384*(1-2), 184–197.
9. Bushueva, T. L., Busel, E. P., & Burstein, E. A. (1978). Relationship of thermal □uenching of protein fluorescence to intramolecular structural mobility. *Biochimica et Biophysica Acta (BBA)-Protein Structure, 534*(1), 141–152.
10. Butzler, T. M., Bailey, J., & Beute, M. K. (1998). Integrated management of *Sclerotinia* blight in peanut: Utilizing canopy morphology, mechanical pruning, and fungicide timing. *Plant Disease, 82*(12), 1312–1318.
11. Carvajal-Yepes, M., Cardwell, K., Nelson, A., Garrett, K. A., Giovani, B., Saunders, D. G., & Tohme, J. (2019). A global surveillance system for crop diseases. *Science, 364*(6447), 1237–1239.

12. Castro, A. I., Ehsani, R., Ploetz, R. C., Crane, J. H., & Buchanon, S. (2015). Detection of laurel wilt disease in avocado using low altitude aerial imaging. *PloS One, 10*(4), e0124642.
13. Chaerle, L., Leinonen, I., Jones, H. G., & Van Der Straeten, D. (2007). Monitoring and screening plant populations with combined thermal and chlorophyll fluorescence imaging. *Journal of Experimental Botany, 58*(4), 773–784.
14. Cock, M. J. W., Tang, R., Liu, Z., Wan, H. H., McGillivray, L. A., Thomas, S. E., & Zhang, F. (2016). The main agricultural insect and disease pests of China and implications for the use of remote sensing for their management. *CABI Reviews, (2016),* 1–23.
15. Cook, R. J., Bruckart, W. L., Coulson, J. R., Goettel, M. S., Humber, R. A., Lumsden, R. D., & Vaughn, J. L. (1996). Safety of microorganisms intended for pest and plant disease control: A framework for scientific evaluation. *Biological Control, 7*(3), 333–351.
16. Cusack, R., & Papadakis, N. (2002). New robust 3-D phase unwrapping algorithms: Application to magnetic field mapping and undistorting echoplanar images. *NeuroImage, 16*(3), 754–764.
17. Delalieux, S., Van Aardt, J. A. N., Keulemans, W., Schrevens, E., & Coppin, P. (2007). Detection of biotic stress (*Venturia inaequalis*) in apple trees using hyperspectral data: Non-parametric statistical approaches and physiological implications. *European Journal of Agronomy, 27*(1), 130–143.
18. Dotto, A. C., Dalmolin, R. S. D., Grunwald, S., Caten, A., & Pereira Filho, W. (2017). Two preprocessing techniques to reduce model covariables in soil property predictions by Vis-NIR spectroscopy. *Soil and Tillage Research, 172,* 59–68.
19. Falkenberg, H. K., Rubin, G. S., & Bex, P. J. (2007). Acuity, crowding, reading, and fixation stability. *Vision Research, 47*(1), 126–135.
20. Ferentinos, K. P. (2018). Deep learning models for plant disease detection and diagnosis. *Computers and Electronics in Agriculture, 145,* 311–318.
21. Frank, L. A., Sigwarth, J. B., Craven, J. D., Cravens, J. P., Dolan, J. S., Dvorsky, M. R., & Muller, D. W. (1995). The visible imaging system (VIS) for the Polar spacecraft. *Space Science Reviews, 71,* 297–328.
22. Gogoi, N. K., Deka, B., & Bora, L. C. (2018). Remote sensing and its use in detection and monitoring plant diseases: A review. *Agricultural Reviews, 39*(4), 307–313.
23. Goodwin, V. A., Richards, S. H., Taylor, R. S., Taylor, A. H., & Campbell, J. L. (2008). The effectiveness of exercise interventions for people with

Parkinson's disease: A systematic review and meta - analysis. *Movement Disorders, 23*(5), 631–640.

24. Gullino, M. L., Albajes, R., Al-Jboory, I., Angelotti, F., Chakraborty, S., Garrett, K. A., & Stephenson, T. (2022). Climate change and pathways used by pests as challenges to plant health in agriculture and forestry. *Sustainability, 14*(19), 12421.
25. Hahn, O., Porciani, C., Dekel, A., & Carollo, C. M. (2009). Tidal effects and the environment dependence of halo assembly. *Monthly Notices of the Royal Astronomical Society, 398*(4), 1742–1756.
26. Hassan, M., Kowalska, A., & Ashraf, H. (2023). Advances in deep learning algorithms for agricultural monitoring and management. *Applied Research in Artificial Intelligence and Cloud Computing, 6*(1), 68–88.
27. Heylen, R., Burazerovic, D., & Scheunders, P. (2011). Fully constrained least squares spectral unmixing by simplex projection. *IEEE Transactions on Geoscience and Remote Sensing, 49*(11), 4112–4122.
28. Homer, C. G., Ramsey, R. D., Edwards Jr, T. C., & Falconer, A. (1997). Landscape cover-type modeling using a multi-scene thematic mapper mosaic. *Photogrammetric Engineering and Remote Sensing, 63*(1), 59–67.
29. Huang, X., Kurata, N., Wei, X., Wang, Z. X., Wang, A., Zhao, Q., & Han, B. (2012). A map of rice genome variation reveals the origin of cultivated rice. *Nature, 490*(7421), 497–501.
30. Iqbal, J., Vogt, M., & Bajorath, J. (2020). Activity landscape image analysis using convolutional neural networks. *Journal of Cheminformatics, 12,* 1–15.
31. Janiesch, C., Zschech, P., & Heinrich, K. (2021). Machine learning and deep learning. *Electronic Markets, 31*(3), 685–695.
32. Jaque, D., Rosal, B. D., Rodríguez, E. M., Maestro, L. M., Haro-Gonzalez, P., & Solé, J. G. (2014). Fluorescent nanothermometers for intracellular thermal sensing. *Nanomedicine, 9*(7), 1047–1062.
33. Jeanneret, P., Schüpbach, B., & Luka, H. (2003). Quantifying the impact of landscape and habitat features on biodiversity in cultivated landscapes. *Agriculture, Ecosystems & Environment, 98*(1-3), 311–320.
34. Keshava, N., & Mustard, J. F. (2002). Spectral unmixing. *IEEE Signal Processing Magazine, 19*(1), 44–57.
35. Kumar, S. (2014). Plant disease management in India: Advances and challenges. *African Journal of Agricultural Research, 9*(15), 1207–1217.

36. Kushwaha, S. P. S., Dwivedi, R. S., & Rao, B. R. M. (2000). Evaluation of various digital image processing techniques for detection of coastal wetlands using ERS-1 SAR data. *International Journal of Remote Sensing, 21*(3), 565–579.

37. Laliberte, A. S., Rango, A., Herrick, J. E., Fredrickson, E. L., & Burkett, L. (2007). An object-based image analysis approach for determining fractional cover of senescent and green vegetation with digital plot photography. *Journal of Arid Environments, 69*(1), 1–14.

38. Lamichhane, J. R., Aubertot, J. N., Begg, G., Birch, A. N. E., Boonekamp, P., Dachbrodt-Saaydeh, S., & Messéan, A. (2016). Networking of integrated pest management: A powerful approach to address common challenges in agriculture. *Crop Protection, 89,* 139–151.

39. Li, W. (2023). Texture feature extraction of a landscape design image based on the contour wave transform. *International Journal of Data Science, 8*(1), 39–51.

40. Luedeling, E., Zhang, M., & Girvetz, E. H. (2009). Climatic changes lead to declining winter chill for fruit and nut trees in California during 1950–2099. *PloS One, 4*(7), e6166.

41. Luo, J., Li, H., Liu, T., Polle, A., Peng, C., & Luo, Z. B. (2013). Nitrogen metabolism of two contrasting poplar species during acclimation to limiting nitrogen availability. *Journal of Experimental Botany, 64*(14), 4207–4224.

42. Luvisi, A., Ampatzidis, Y. G., & De Bellis, L. (2016). Plant pathology and information technology: Opportunity for management of disease outbreak and applications in regulation frameworks. *Sustainability, 8*(8), 831.

43. MacDougall, S., Bayansal, F., & Ahmadi, A. (2022). Emerging methods of monitoring volatile organic compounds for detection of plant pests and disease. *Biosensors, 12*(4), 239.

44. Mahlein, A. K. (2016). Plant disease detection by imaging sensors–parallels and specific demands for precision agriculture and plant phenotyping. *Plant Disease, 100*(2), 241–251.

45. Martinelli, F., Scalenghe, R., Davino, S., Panno, S., Scuderi, G., Ruisi, P., & Dandekar, A. M. (2015). Advanced methods of plant disease detection. A review. *Agronomy for Sustainable Development, 35,* 1–25.

46. Naidu, M. M., Shyamala, B. N., Naik, J. P., Sulochanamma, G., & Srinivas, P. (2011). Chemical composition and antioxidant activity of the husk and endosperm of fenugreek seeds. *LWT-Food Science and Technology, 44*(2), 451–456.

47. Naskath, J., Sivakamasundari, G., & Begum, A. A. S. (2023). A study on different deep learning algorithms used in deep neural nets: MLP SOM and DBN. *Wireless Personal Communications, 128*(4), 2913–2936.
48. Nilsson, O., Wängberg, B., Kölby, L., Schultz, G. S., & Ahlman, H. (1995). Expression of transforming growth factor alpha and its receptor in human neuroendocrine tumors. *International Journal of Cancer, 60*(5), 645–651.
49. Permyakov, E. A., & Burstein, E. A. (1984). Some aspects of studies of thermal transitions in proteins by means of their intrinsic fluorescence. *Biophysical Chemistry, 19*(3), 265–271.
50. Prabhakar, M., Prasad, Y. G., Thirupathi, M., Sreedevi, G., Dharajothi, B., & Venkateswarlu, B. (2011). Use of ground-based hyperspectral remote sensing for detection of stress in cotton caused by leafhoppers (*Hemiptera: Cicadellidae*). *Computers and Electronics in Agriculture, 79*(2), 189–198.
51. Prabhakar, M., Thirupathi, M., & Mani, M. (2022). Principles and application of remote sensing in crop pest management. *Trends in Horticultural Entomology,* 157–183.
52. Pu, H., Xia, W., Wang, B., & Jiang, G. M. (2013). A fully constrained linear spectral unmixing algorithm based on distance geometry. *IEEE Transactions on Geoscience and Remote Sensing, 52*(2), 1157–1176.
53. Pydipati, R., Burks, T. F., & Lee, W. S. (2006). Identification of citrus disease using color texture features and discriminant analysis. *Computers and Electronics in Agriculture, 52*(1-2), 49–59.
54. Ramirez-Lopez, L., Behrens, T., Schmidt, K., Stevens, A., Demattê, J. A. M., & Scholten, T. (2013). The spectrum-based learner: A new local approach for modeling soil Vis-NIR spectra of complex datasets. *Geoderma, 195,* 268–279.
55. Ratnadass, A., Fernandes, P., Avelino, J., & Habib, R. (2012). Plant species diversity for sustainable management of crop pests and diseases in agroecosystems: A review. *Agronomy for Sustainable Development, 32,* 273–303.
56. Regmi, N. R., Giardino, J. R., McDonald, E. V., & Vitek, J. D. (2014). A comparison of logistic regression-based models of susceptibility to landslides in western Colorado, USA. *Landslides, 11,* 247–262.
57. Riley, M. (1989). Some applications of tree-based modeling to speech and language. In *Speech and Natural Language: Proceedings of a Workshop Held at Cape Cod, Massachusetts, October 15-18, 1989* (pp. 100–109).
58. Römer, C., Bürling, K., Hunsche, M., Rumpf, T., Noga, G., & Plümer, L. (2011). Robust fitting of fluorescence spectra for pre-symptomatic

wheat leaf rust detection with support vector machines. *Computers and Electronics in Agriculture, 79*(2), 180–188.

59. Römer, I., White, T. A., Baalousha, M., Chipman, K., Viant, M. R., & Lead, J. R. (2011). Aggregation and dispersion of silver nanoparticles in exposure media for aquatic toxicity tests. *Journal of Chromatography A, 1218*(27), 4226–4233.
60. Rossel, R. V., Cattle, S. R., Ortega, A., & Fouad, Y. (2009). In situ measurements of soil color, mineral composition, and clay content by Vis-NIR spectroscopy. *Geoderma, 150*(3-4), 253–266.
61. Salehin, I., Noman, S. M., Baki-Ul-Islam, I. J., Bishnu, P., Habiba, U., & Nessa, N. (2021). IFSG: Intelligence agriculture crop-pest detection system using IoT automation system. *Indonesian Journal of Electrical Engineering and Computer Science, 24*(2), 1091–1099.
62. Silva, G., Tomlinson, J., Onkokesung, N., Sommer, S., Mrisho, L., Legg, J., & Boonham, N. (2021). Plant pest surveillance: From satellites to molecules. *Emerging Topics in Life Sciences, 5*(2), 275–287.
63. Singh, R. J., Chung, G. H., & Nelson, R. L. (2007). Landmark research in legumes. *Genome, 50*(6), 525–537.
64. Southwood, T. R. (1977). Habitat, the templet for ecological strategies? *Journal of Animal Ecology, 46*(2), 337–365.
65. Steddom, K., Heidel, G., Jones, D., & Rush, C. M. (2003). Remote detection of rhizomania in sugar beets. *Phytopathology, 93*(6), 720–726.
66. Stenberg, J. A. (2017). A conceptual framework for integrated pest management. *Trends in Plant Science, 22*(9), 759–769.
67. Terentev, A., Dolzhenko, V., Fedotov, A., & Eremenko, D. (2022). Current state of hyperspectral remote sensing for early plant disease detection: A review. *Sensors, 22*(3), 757.
68. Vernon, D., & Sandini, G. (1988). VIS: A virtual image system for image understanding research. *Software: Practice and Experience, 18*(5), 395–414.
69. Willby, N. J., Abernethy, V. J., & Demars, B. O. (2000). Attribute based classification of European hydrophytes and its relationship to habitat utilization. *Freshwater Biology, 43*(1), 43–74.
70. Witoszynskyj, S., Rauscher, A., Reichenbach, J. R., & Barth, M. (2009). Phase unwrapping of MR images using ΦUN–A fast and robust region-growing algorithm. *Medical Image Analysis, 13*(2), 257–268.
71. Xu, W., Zhang, G., Li, X., Zou, S., Li, P., Hu, Z., & Li, J. (2007). Occurrence and elimination of antibiotics at four sewage treatment plants in the Pearl

River Delta (PRD), South China. *Water Research, 41*(19), 4526–4534.

72. Yamauchi, M., Tokeshi, M., Yamaguchi, J., Fukuzawa, T., Hattori, A., Hibara, A., & Kitamori, T. (2006). Miniaturized thermal lens and fluorescence detection system for microchemical chips. *Journal of Chromatography A, 1106*(1-2), 89–93.
73. Yang, X., Yan, J., Shah, T., Warburton, M. L., Li, Q., Li, L., & Li, J. (2010). Genetic analysis and characterization of a new maize association mapping panel for quantitative trait loci dissection. *Theoretical and Applied Genetics, 121,* 417–431.
74. Yao, J., Wang, L., Liu, L., Zhao, C., & Zheng, Y. (2009). Association mapping of agronomic traits on chromosome 2A of wheat. *Genetica, 137,* 67–75.
75. Yuan, Y., Zheng, X., & Lu, X. (2017). Hyperspectral image superresolution by transfer learning. *IEEE Journal of Selected Topics in Applied Earth Observations and Remote Sensing, 10*(5), 1963–1974.
76. Zhang, X., Davidson, E. A., Mauzerall, D. L., Searchinger, T. D., Dumas, P., & Shen, Y. (2015). Managing nitrogen for sustainable development. *Nature, 528*(7580), 51–59.

INDEX

C

D

E

F

G

H

R

S

T

FSC
www.fsc.org
MIX
Papier | Fördert gute Waldnutzung
FSC® C083411

Kolibri 360 GmbH
Ferdinand-Jühlke-Straße 7
99095 Erfurt, Deutschland
produktsicherheit@kolibri360.de